What Is Sustainability?

What Is Sustainability?

*An Overview of Our Impact on Planet Earth
and the Natural Forces Shaping our Future*

什么是可持续发展?

IAN SPELLERBERG, *Editor*

BERKSHIRE PUBLISHING GROUP

Great Barrington, Massachusetts

Permissions may also be obtained via Copyright Clearance Center, 222 Rosewood Drive, Danvers, MA 01923, USA, telephone +1 978 750 8400, fax +1 978 646 8600, info@copyright.com.

Digital editions. *What Is Sustainability?* is available through most major e-book and database services (please check with them for pricing).

For information, contact

Berkshire Publishing Group
122 Castle Street
Great Barrington, Massachusetts 01230-1506 USA
Email: info@berkshirepublishing.com
Tel: +1 413 528 0206
Fax: +1 413 541 0076

Cover photograph by Carl Kurtz.

Library of Congress Cataloging-in-Publication Data
Names: Spellerberg, Ian F., editor.
Title: What is sustainability? : an overview of our impact on planet earth and the natural forces shaping our future / Ian Spellerberg, Editor.
Description: Great Barrington : Berkshire Publishing Group, 2020. | Includes bibliographical references and index.
Identifiers: LCCN 2019048211 | ISBN 9781933782867 (paperback)
Subjects: LCSH: Sustainability. | Sustainable development. | Nature—Effect of human beings on. | Global environmental change.
Classification: LCC HC79.E5 W475 2020 | DDC 304.8—dc23
LC record available at https://lccn.loc.gov/2019048211

Table of Contents

Introduction

The field of environmental studies these days is almost synonymous with controversy. Whatever the topic—climate change, genetically modified organisms, fracking, the Green Revolution—there are many different positions, with proponents who claim that their views are supported by science, politics, and history.

The history of environmental studies goes back to the eighteenth century. Globally, a very important milestone for environmental studies was the 1972 United Nations Conference on the Human Environment held in Stockholm, Sweden. In effect, this was the first of what was to become an Earth Summit every ten years. The Conference agreed upon a declaration containing twenty-six principles concerning the environment and development, an Action Plan with 109 recommendations, and a resolution. One of the many initiatives that took place afterwards was the establishment of interdisciplinary environmental studies. The emphasis was on "interdisciplinary," whereby students were taught how several disciplines could be combined to address environmental issues.

Graduates from those interdisciplinary environmental studies programs were initially treated with some suspicion by potential employers because they were seen as generalists rather than specialists. That perception soon changed, as did the perception of sustainability. In effect, those interdisciplinary programs helped to define the aim of environmental studies: how to achieve a sustainable and equitable use of nature (the living components) and the environment (the non-living components). Sustainability is therefore sufficiently broad and inclusive. It provides a way to identify, appreciate, and measure changes, and it makes connections between environmental issues and other global challenges.

One useful way to look at sustainability is to consider nature and environment as sources and "sinks" used by humans. The environment provides sources of food, fiber, minerals and shelter, etc. Humans also use the environment as a sink to absorb solid, liquid and gaseous wastes. Both sources and sinks have limits. For examples, excessive use of agricultural soils leads to degradation of those soils. Put too much pollution into the atmospheric sink and the environment's ability to absorb those pollutants is exceeded. Fishing becomes unsustainable when the rate of removal of fish is faster than those fish populations can replenish themselves.

The concept of sustainability is all about human-environment relationships and the extent to which humans live within environmental limits. Unfortunately, since 1972, much of humanity has been exploiting nature and the environment in an increasingly unsustainable and inequitable manner.

The purpose of *What Is Sustainability?* is to set forth the issues concerning the human-environment relationship, showing the various sides of these issues, so that those entering the vast world of environmental studies can begin to understand the breadth of this field. One way to look at the environment that may make it less abstract is to look at it as a community: a home shared by all. After all, the word *ecology*, which today refers to the branch of biology that studies the interrelationships between organisms and the environment, comes from the Greek *oikos* and means "the study of homes."

Underlying our approach is the belief that this understanding is best achieved by considering what both scientists and historians have to tell us. Science is a method (although not the only one, of course) for learning about the world. By *science* we mean not only the physical sciences but also the social sciences, because we recognize that environmental discussions take place in a social, economic, and political context. History is a method of learning about the past to inform the present. Humans (or our early ancestors) have been altering the environment for at least 4 million years, so we must learn from the past.

We live in a world facing serious issues, and at least a basic knowledge of them is essential to becoming a global citizen. *What Is Sustainability?* covers the human-environment relationship in six chapters organized into logical and manageable topics. The first two chapters focus on prehistory and history, from the earliest human ancestors through the millions of years of hunting and gathering, continuing through the rise and spread of agriculture and the creation of a connected, global environment. In Chapter 3, we move on to the industrial, urban, modern world and the merging of old and new human impacts on the environment. Chapters 4 and 5 cover the major environmental issues of our time—resource overuse, energy, agriculture, and a growing population, to name a few—as well as solutions and proposed solutions within the sustainability framework. Although this book is primarily about history, Chapter 6 ventures into the future. We end with a postscript about the current situation and changes we can all make in our lives, followed with a list of sources and suggestions for further reading for each chapter. Let's start at the beginning.

CHAPTER 1

Humanity's First Steps

Time present and time past
Are both perhaps present in time future,
And time future contained in time past.

—*T.S. Eliot (1888–1965)*

Astronomers and geologists generally concur that the Earth is between 4.5 and 4.8 billion years old. They have determined this by analyzing ancient rocks to find the rate at which uranium decays into lead, as well as by measuring the age of meteorites and moon rocks.

Some 300 to 250 million years ago, the world's continents fused to form a single supercontinent, called Pangaea. Creatures formerly kept apart now came into contact with each other, and large numbers of them went extinct by about 220 million years ago. Reptiles inherited the Earth, spreading throughout the globe.

The geological time chart—a system for dating events in the Earth's history based on rock stratification—contains several different types of time periods. The largest are the *eons*, such as the Phanerozoic, the era of large organisms, which covers the last 540 million years. The next largest are the *eras*, such as the Cenozoic, the era of mammals, which covers the last 66 million years. Eras, in turn, can be divided into *periods*, such as the Quaternary, which covers the last 2 million years. Finally, *epochs* subdivide periods. The last and shortest of the epochs is the Holocene, which includes the 11,500 years since the end of the last Ice Age, a period of unusual climatic stability.

A number of scholars argue that the Holocene has ended, claiming that in the last two centuries we have in fact entered a new epoch, dubbed the *Anthropocene* (from Greek roots meaning "human" and "new"), a turbulent period of exceptional and unpredictable change. As the name suggests, the defining feature of the Anthropocene is the transformative role played by our own species, *Homo sapiens*. For most modern humans living during this time, increasing human control over the biosphere—the sliver of the Earth's crust that supports life—has meant a vast improvement in living standards, better communications, and faster transportation.

In the last fifty years, however, it has become apparent that these gains may have come at a considerable cost to the world environmental community, and that we may be provoking what many observers call the sixth great extinction spasm in the history of the Earth. Humankind has become a global geological force in its own right, but the notion that we might have become the dominant force for change in the biosphere emerged only in the last two decades of the twentieth century.

The Anthropocene is noteworthy even on the huge time scales of planetary history, because it marks the first time in the almost 4-billion-year history of life on Earth that a single species has played the leading role in shaping the biosphere. We are changing things rapidly. Never before has a single species had the power to transform the entire biosphere in just a few centuries.

A New Geological Epoch

Although the Anthropocene epoch and the changes associated with it may seem recent, they are the culmination of processes that go back to the beginnings of human history. Our technological precocity as a species already was apparent in the artistic and technological skills evident at archaeological sites such as Blombos Cave in South Africa. Here, almost 100,000 years ago, humans learned how to use shellfish, developed sophisticated color palettes, and carved intricate patterns on ocher rocks.

Global migrations showed our remarkable technological creativity. From about 60,000 years ago, small communities of humans moved to all of the world's continents except Antarctica. Each migration required new strategies and technologies to deal with unfamiliar climates, plants, animals, and physical environments. The technological revolutions that have created the Anthropocene epoch in the last two centuries represent a sharp acceleration in processes that are as old as *Homo sapiens*.

The evidence that we have entered a different geological epoch is extensive and varied, including: a tenfold increase in human population since 1700, human exploitation of up to 50 percent of the Earth's surface, increasing human control of water flows, and human control of more than 50 percent of all freshwater, and a sixteenfold increase in human energy use since 1900, which has doubled emissions of sulfur dioxide.

Scientists point out two pieces of evidence: the disruption of the nitrogen cycle, and the phenomenon known as the "human appropriation of net primary production," or HANPP. As for the nitrogen cycle, commercial agriculture relies heavily on industrial nitrogen fixation, which puts more nitrogen compounds into the environment than would naturally occur, enabling greater agricultural productivity but causing environmental harm. These anthropogenic sources of nitrogen (industrial nitrogen fixation, the use of leguminous crops, fossil fuels, etc.) impact the nitrogen cycle. As for HANPP, agriculture and other human uses of the land alter stocks and flows of biomass in ecosystems. HANPP accounts provide information both on the scale of human activities compared to natural flows and on the aggregate intensity of land use.

Directly or indirectly, the thread that unites these and other changes is the dynamism of economies and communities based on the rapidly increasing use of fossil fuels. Earlier human societies could access only limited reserves of energy, derived from recently captured solar energy. That energy was tapped either as wind or water power (as solar energy moved the planet's air and water masses), or in the form of foods and fuel energy from plants (solar energy captured through photosynthesis), or through the food and energy supplied by domesticated animals and slaves (which represented solar energy captured by plants and recaptured by animals and humans higher up the food chain). All these sources of energy ultimately derived from solar energy that had been captured within recent decades or centuries.

In contrast, fossil fuels—coal, oil, and natural gas—represent flows of solar energy captured and buried over several hundred million years. Scientists estimate that human energy use has increased by at least forty times since 1800, and today fossil fuels supply at least 85 percent of human energy. With the help of what seemed to be an inexhaustible supply of energy, humans found they could mobilize and control the resources of the biosphere on unimaginable scales. These unprecedented energy and resource flows helped human populations rise from 1 billion to 7 billion in two hundred years, while global production of goods and services increased at least fiftyfold.

The vast increase in human control of biospheric resources has become apparent since the middle of the twentieth century, when, after a period of slower growth and destructive global wars, growth took off faster than ever before. This was driven in part by wartime innovations, such as nuclear power and computers. The sharp acceleration after 1950 marks a second phase in the Anthropocene. The changes made possible directly or indirectly by the fossil fuels revolution explain why our species now plays such a decisive role in the biosphere, and why so many scholars date the Anthropocene epoch to the Industrial Revolution and the invention of an improved steam engine.

The scale of these changes is apparent in all key sectors of the biosphere. The burning of fossil fuels has significantly raised levels in the atmosphere of greenhouse gases such as carbon dioxide (CO_2) and methane. There is growing evidence that this change is raising global temperatures, transforming climates across the world, and causing a significant rise in sea levels as oceans expand and ice sheets melt. The stability of CO_2 ensures that once it is in the atmosphere, such changes will take many centuries to reverse. Less predictable, but perhaps more worrying, is the possibility that there may be dangerous tipping points beyond which climate could change very quickly indeed.

Burning fossil fuels alters the hydrosphere, which is all of the water found in the atmosphere and on and under the planet's surface. Absorption of CO_2 in the world's oceans is temporarily slowing the rise in atmospheric CO_2. The increasing acidity levels in the oceans, however, affects coral systems and the many other marine organisms that produce calcium to make their shells or skeletons. Humans are transforming the land through construction, mining, dam construction, and agriculture. Although the situation is dire indeed, our best hope for the future comes from concerted action by the human community.

Biodiversity is decreasing as rates of extinction approach levels seen only during the greatest mass extinctions of the last 600 million years. By some estimates, extinction rates are between one hundred and one thousand times the "background level" for the last few million years. The primary causes are overfishing the oceans, and human transformation of the land, the latter of which is destroying or breaking up habitats. Other human activities also contribute, such as the disposal of plastics, waste products, and other pollutants in the land and in the oceans.

Natural cycles are disrupted as the burning of fossil fuels in the last century has transformed the carbon cycle by pumping into the atmosphere carbon that was buried over several hundred million years. Meanwhile, the burning of fossil fuels and the manufacture of artificial fertilizers is converting vast amounts of nitrogen into more reactive forms, and doing so at a speed and on a scale that can no longer be managed by the natural nitrogen cycle.

One of the most spectacular measures of increasing human power within the biosphere is the evolution of nuclear weapons, which have given humans the power, in principle, to destroy much of the biosphere in just a few hours.

A recent attempt to identify when changes such as these crossed dangerous thresholds concludes that in three areas—carbon dioxide levels, flows of nitrogen, and rates of extinctions—we have already moved beyond safe "planetary boundaries."

What It Means for Our Future

Such changes raise profound questions about the nature and future of our own species. Why are humans so creative? The best answer seems to be that human language crossed a tipping point that allowed such precise information to be exchanged that it could be stored within the collective memory and accumulate from generation to generation. That increasing stock of information is the ultimate source of humanity's increasing control over the biosphere. We are probably the first such species in the history of our planet, because any earlier species with such powers would have left paleontological traces as clear as those we will leave to future generations.

As the Dutch climatologist Paul Crutzen has argued, the idea of the Anthropocene is a call to action because it points to dangers that need to be tackled soon if the human community is to avoid serious—and perhaps catastrophic—changes in coming decades. The problem is partly a matter of speed. Human-induced changes, both within human societies and in the biosphere as a whole, are now proceeding so fast that natural processes cannot keep up. We humans struggle to keep track of the many changes we are causing. Human activity is pushing the biosphere into uncharted territory.

The terrifying pace and scale of change in the Anthropocene suggests that we need to slow those changes having the greatest impacts on the biosphere. This would be wise not because we *know* where change is leading, but because we do *not* know where it is leading. Slowing down means reducing the scale of change, but not slowing the pace of technological change, because new technologies may play a crucial role in reducing human impact on the biosphere. Nor does slowing down mean giving up on the many gains in welfare that are the most positive side of the Anthropocene era.

To fathom the meaning of the Anthropocene era we must first to go back several million years in time to review what we know about the emergence of the era's key actor—human beings—and how we came to inhabit the Earth.

Evolution and the Spread of Humans

Humans comprise one species, *Homo sapiens*, with a wide variety of body forms and physical characteristics. More than 7 billion humans inhabit the Earth, and the population will increase in the foreseeable future. Humans' closest living relatives among the nonhuman primates, the chimpanzees and gorillas, may succumb to the continued growth of humans. Humans differ from these other primates in countless ways, including their manner of walking, their diets, their social structure, and their manifest ability to manipulate their environments.

The first humans were different from their predecessors, the australopithecines, in nearly every anatomical aspect. With statures approaching 1.8 meters, early humans were twice the height of australopithecine females, and female humans were nearly as large as males, unlike the great sexual differences of earlier hominids. Human teeth were smaller, while the brain had expanded from an australopithecine average of 500 cubic centimeters or less to more than 800.

These features reflect a different interaction with the environment than in any earlier hominid. The long limbs of these larger hominids allowed them to range farther from water and allowed them more efficient dissipation of heat. These hominids were probably the first to sweat like living humans, a valuable adaptation to daytime activity in a hot, dry climate. The adaptation to sweating means that these hominids, like humans, were largely hairless.

The division of labor in food acquisition between the sexes, which characterizes modern human-foraging groups, may have originated at this time. Hunting prey or scavenging the prey of other carnivores may have been an important part of food acquisition. Still, as with foragers in modern times, plant foods probably made up most of the diet. These social changes may have gone hand in hand with a prolonged period of child development. The brains of early humans were within the lower range of those of living people and are the first to exhibit features, such as left-right asymmetry, common to humans. The increases in technology that stone tools reflect were certainly part of a trend toward larger brains. These changes are, in fact, the first signs of the development of culture. These hominids did not use a language quite like that of humans, but language had begun to evolve, as changes in the vocal tract and brain structure show. These hominids—the first that people would recognize as human—depended on one another, as we still do today.

Colonization of the "Old World"

Almost immediately after their origin, early humans moved out of Africa and spread across large parts of Eurasia, commonly known as the "Old World" to distinguish it from the "New World" of the Americas. The earliest sites documenting this spread are in the Caucasus, the eastern Mediterranean near present-day Lebanon, Java, and China. Most of the subtropical Old World appears to have been occupied by 1 million years ago. This spread appears to be directly associated with the new human features and ways of life, because no australopithecines lived outside of Africa. Despite their widespread distribution, humans were much more limited in behavioral variation than are today's humans. A few tool industries, with large-scale regional differences, existed between 1.8 million and 400,000 years ago. These techniques persisted over a long time, and it is clear that the rate of social and technological changes that characterizes living humans was almost nonexistent among early humans.

Despite the spread of many people, Africa remained the center of the human population. It was in Africa that the first humans evolved and in Africa that they would remain at the highest density until recent times. Large numbers of people meant that African populations contained more genetic diversity than other regions. With fewer people, populations in other parts of the Old World began to diversify.

In the view of many scientists, these regional variations proceeded so far that different species of humans began to appear. The regions of the Old World became home to populations of humans who did not interbreed with one another. Only one of these ancient species was ancestral to today's humans, according to this view, with other species going extinct during the late Pleistocene epoch as a result of competition between them.

Many geneticists believe that modern humans' direct ancestors numbered only a few thousand people during the past 2 million years. Today's populations began to diverge from this ancestral population only recently, within the past 100,000 years. Africa was likely the homeland of such an ancestral population—this is known as the "Out of Africa" theory.

From their initial spread out of Africa, humans across the world began to change. The most notable change was the increase in brain size: from a beginning size of around 800 milliliters to the worldwide average of around 1,350 milliliters today. The changes in brain size triggered many related changes in the skull.

Around 200,000 years ago, technologies began to change as regional variations in tool use became important for the first time. People in different areas of Africa experimented with diverse tools, many of which were later discarded. Across Europe and Western Asia, new stone tools appeared, created by what is called the *Levallois technique*, named for a suburb of Paris. This complex technique, which results in a well-formed tool created after a series of steps that seem to have nothing to do with the tool's form, is thought by some to be a sign of a modern intelligence (the structured sequence resembles mental activities like language).

Latest Human Evolution

The Middle Paleolithic period began a reordering of the relationship between humans and their environments, one that extended into the succeeding Upper Paleolithic period and up to the present day. The remains of hunted animals show that humans began to exploit a broader range of animal resources. Earlier humans hunted medium-size herbivores, such as sheep, deer, or gazelles, then during the late Pleistocene epoch they added more risky prey, such as Cape buffalo and mammoths. They took easily caught, slow-growing animals, such as land tortoises and some shellfish, in large quantities.

There is evidence that humans in Australia (where migrants first arrived earlier than 60,000 years ago) burned grasslands as early as 50,000 years ago, to attract animals to new growth. Across the Old World and later the New World, the Pleistocene megafauna—from the rhino-sized marsupials of Australia to the mammoths of Europe—began to disappear. Human population increases and consequent stress on other natural populations likely caused these shifts.

At the same time, humans increased their technological abilities to allow the exploitation of other resources. String had been invented in Europe by around 25,000 years ago (judging by the lines left in pottery shards), and humans began to make nets and snares for fish and small animals. This change, which spread the ability to acquire animal protein, appears to have accompanied an increasingly settled lifestyle of human populations as population pressures began to prohibit the free movement of peoples and to encourage the year-round exploitation of seasonal resources (as indicated by the increasing use of food storage pits).

Larger populations may have caused increases in cultural differentiation, with social networks extending beyond the numbers of people that any individual might know well. Cultural ornaments and artwork, which may have communicated status or other social information, became much more common during Upper Paleolithic times. The artworks gracing caves such as Lascaux in France, as well as Australian rock faces that are even older, use persistent symbols, reflecting a concern with communicating with people who are distant in place or in time. These symbols, along with the elaborate human burials that characterized Middle as well as Upper Paleolithic peoples, are among the earliest signs of a human concern with the nonmaterial.

Human evolution continues to occur. As humans alter their environments, they are placed in novel ecological circumstances that have reproductive and evolutionary consequences. Humans did not originate in social groups of thousands of people. The large size of the current human population implies that some evolutionary changes, such as the natural selection that adapts humans to their environment, will occur only slowly in the future.

Furthermore, human cultural practices, including medicine, environmental alteration, and economic systems, often reduce the genetic relationship between human behavior or physical form and their reproduction. Other effects of the new human population structure may be accelerating, however. Today people move between populations around the world with increasing ease, and as a result human genes are mixing at an increasing rate. Humans have mixed the world's biogeography—that is, the past and present distribution of the many species found on our planet. Large demographic differences between societies exist, with the population growth rate different between regions. The human population of the future will have a genetic makeup different from those people alive today. Whatever the genetic makeup of humans beings, it is clear that the number of human beings on Earth, where they live, and how they interact with their environments influence the environment in numerous ways, a subject we address in the next chapter.

CHAPTER 2

From Foraging to Growing

Agriculture represents the single most profound ecological change in
the entire 3.5 billion-year history of life.

—Niles Eldredge (1943–)

Perhaps the key technological change in the human experience was
the transformation from foraging for food to cultivating and growing
food. This will be the focus of this chapter. We begin with the subsistence strategy—hunting and gathering—that sustained human communities for all but about the last 12,000 years of our existence, and then follow
with agriculture. Agriculture replaced hunting and gathering all over the
Earth, and we examine it here in its various forms from subsistence gardening to modern agribusiness. This chapter concludes with a broad examina-
tion of biological exchanges: the movement of organisms from one region to
another.

Many species became extinct during the late Pleistocene/early Holocene period, a time when not only did the climate change rapidly, but the
first humans arrived. Evidence suggests that humans were at least somewhat
responsible for the extinctions of this period. North America and Australia,
in particular, provide stages for scientists' reenactment of the disappearance
of many large herbivorous mammal species and genera, together with their
dependent carnivores.

Hunting and Gathering

The history of the hunting and gathering (also called foraging) way of life covers about 4 million years. It begins with the time of emerging humans, covers

the millennia when hunter-gatherers constituted all of the human species, and continues through their large-scale disappearance in the face of agriculture into the twentieth century. Groups of hunter-gatherers still exist, although they no longer subsist by foraging alone and often trade with their neighbors.

Hunting-gathering, which also encompasses fishing and scavenging, is a mode of subsistence characterized by the absence of direct human control over the reproduction of exploited species, with the exception of the dog. It is possible, however, for hunter-gatherer groups to bring about environmental change. Sometimes this appears deliberate in its attempt to improve the conditions for resource species, and at other times it may well be accidental.

Regions from the lowland tropical forests to the high Arctic, except for those permanently covered in ice and the deep oceans far from land, have housed a population of hunter-gatherers at some stage during the Pleistocene (beginning 1.6 million years ago) or Holocene (beginning 10,000 years ago) epoch. Their environmental relationships were characterized by slow population growth and few material possessions. In the broadest of terms, the degree of dependence upon plant foods declines away from the equator, and there is a reciprocal relationship between the concentration of calories and the concentration of their source. Given the social and ecological adaptability of most groups, a satisfactory diet was often achieved without the sort of backbreaking labor required by subsistence agriculture. The price was, usually, intermittent famines.

Hunter-gatherer communities disappeared after agriculture was irreversibly established, except in a few areas marginal for agriculture or at least remote from it. The culture of descendants of ancient hunter-gatherers is framed by their lineage: the Kalahari San people were nonagricultural from the late Stone Age (8,000 years ago) until the late nineteenth century.

Fishing is a more recent development in human adaptations and is often considered to be a specialized form of hunting and gathering. Fishing is a distinct form of subsistence, especially when it involves the taking of large quantities of fish to be preserved for year-round use. This type of fishing, used by peoples such as Native Americans in the Pacific Northwest of the United States, allowed for a sedentary and often rich lifestyle, leading these types of fishing societies to be labeled "affluent hunter-gatherers."

The first clear evidence of fishing points to roughly 32,000 to 25,000 years ago, but the rarity and sporadic distribution of evidence indicate only low-level exploitation of fish such as salmon during the late-upper Paleolithic period in western Europe. It was not until after the Pleistocene-Holocene epoch transition 10,000 years ago, however, that fishing became widespread along coastlines, rivers, lakes, and swamps worldwide. Humans' means for capturing fish diversified rapidly, including larger watercraft, nets, and complex lines and weirs (fences or enclosures set in a waterway for taking fish) that facilitated the capture of numerous and/or larger fish.

One issue of great importance in the human-nature relationship from early times onward is the possession and control of fire. Fire at the hearth widens the dietary possibilities of hunter-gatherers because it breaks down some foods that the human gut has trouble digesting. More significantly, fire allows the drying and smoking of meat and fish as methods of preservation. Humans had the ability to control fire at the landscape scale. It seems certain that Homo sapiens possessed this skill, but for earlier hominid species the evidence is indirect. Most archaeologists, however, believe that such a proficiency was part of the tool kit that allowed Homo erectus to spread beyond Africa into the rest of the Old World.

Once mastered, fire has many uses for hunter-gatherers. Fire can be used to herd animals because most are afraid of it and can be steered into the paths of concealed hunters or driven over cliffs and into cul-de-sacs such as box canyons or crescent-shaped dunes. An advancing fire front also puts up an array of small but edible animals: locusts and grasshoppers, small mammals, and confused birds all may be added to the menu, as well as many subterranean lizards and snakes that have been baked. The vegetation that springs up afterward, often after the next rainy season or in the spring in a temperate latitude, is usually richer in protein than unburned stands. Animals then attracted to the plants are concentrated for hunting. Some plants fruit more heavily after fire, and so direct supplies of nuts, for example, are enhanced.

Thus the focus on hunter-gatherers as makers of landscapes during the Holocene epoch after 10,000 years is on their capacity to alter the ecology of a region in a quasi-permanent fashion. These imposed changes include: reductions in some animal and plant populations, conservation of some animals through the voluntary killing of fewer animals, and minimal impact on some types of land such as dry lands.

Many groups were able to exterminate animals on a large scale, and did so sometimes after contact with external traders. One classic example is the penetration of the Hudson's Bay Company into what is now Canada. The company traded in valuable beaver pelts and contributed to a condition in which the beaver harvest in North America in the late nineteenth century was only about 10 percent of its level one hundred years earlier because the animal itself had become scarce or locally extinct.

Hunter-gatherers were not necessarily stewards of the environment, and their subsistence activities often changed the environment, if only in limited ways. The following section on the so-called pristine myth provides one example of how hunter-gatherers altered their environment.

The Pristine Myth

The pristine myth is the belief that until 1492 most of North America and Amazonian South America was a wild, sparsely populated wilderness, full of virgin forests, and teeming with wild animals. The few "primitive" Indians lived so in tune with nature that they barely disturbed their environments. Hollywood, the general media, and schoolbooks perpetuate the myth.

The pristine myth was invented by American nineteenth-century writers such as James Fenimore Cooper, Henry David Thoreau, and Henry Wadsworth Longfellow, who romanticized the American wilderness, the Native Americans who lived there, and the European colonists who settled it. In the mid-1900s, geographers, historians, and anthropologists began questioning the idea that the American environment was pristine at the time of the Genoese explorer Christopher Columbus's arrival in 1492. By the 1980s, new studies in historical ecology had discredited the myth. In the 1990s, scientists presented revisionist arguments that forests in the Americas and tropical forests in Asia and Africa were not virgin, but anthropogenic (human-modified). In the Americas, geographers found that what appeared to be empty lands contained the remains of earthworks, roads, fields, settlements, deforested areas, severe erosion, and wildlife extinction.

The pristine myth envisioned the tropical rainforests of Asia, Africa, and the Amazon to be filled with a cornucopia of wild foods, so much so that prehistoric foraging peoples must have found plenty of nutritious wild foods to eat with almost no labor. By the late 1980s, however, anthropologists presented evidence that such forests are actually so food-poor that most humans could never have lived there without growing some food or trading with neighboring groups who farmed. Since the mid-1990s, most geographers and anthropologists agree that the pristine myth is just that: a myth.

As the largest tropical rainforest on the planet, Amazonia holds a unique place in both world environmental history and in the imagination of South America. A vast region that extends through present-day Brazil and seven other South American nations, it symbolically stands for the dominance of nature over humans, and is a source of still-unknown plants and animals. In fact, however, Amazonia has been a managed, human-made environment for hundreds of years.

Research in ancient Amazonia has focused on anthropic or anthropogenic soils, whose formation is directly related to human activity—or at least on assessing what kind of soils may have been generated through human activities, how widespread they are, and to what extent such soils were formed. The banks of the main channel of the Amazon, as well as of many of its tributaries, are replete with "black earth" sites, illustrating both the continuity and antiquity of human presence. The use of such sites for agricultural purposes thus illustrates both sophisticated knowledge of soil properties and systems of agricultural management that were stable over many generations.

These kinds of soils, particularly *terra preta* (black earth), which is black anthropogenic soil with enhanced fertility due to high levels of soil organic matter and nutrients, are common throughout Amazonia. (Agricultural fertilization or human settlement creates the valuable soil; human by-products such as pottery shards and food scraps enrich the soil with nitrogen.)

Scientific investigation of anthropogenic soils provides evidence that human occupation of an area does not depend on conducive environmental conditions.

Agriculture

Agriculture comprises both cultivation and herding: the human management of populations of plants and animals to produce food and fiber. Agriculture is often defined as food production based on domesticated plants, with or without domesticated animals. Systems based on domesticated animals alone may be included, but are alternatively distinguished as herding or pastoralism.

The accepted criteria of domestication are morphological changes resulting from human control of plant and animal reproduction. Modification of plants and animals is almost unavoidable under human management. For example, harvesting selects for seeds that stay longer on the plant. Maize (corn) is an extreme case, having been so altered that human assistance is needed to disperse the seeds clinging to the mature ears.

From Foraging to Agriculture

Transitions to agriculture probably began with efforts to increase the yields of wild plants, not expressly to make them dependent on human care. Agriculture is a Holocene epoch phenomenon that arose independently in widely separate locations. The earliest domesticated varieties of wheat and barley in Southwest Asia are dated circa 9250 BCE. Sheep, goat, and cattle domestication followed a millennium or more after that elsewhere in the region. Domesticated rice in the Yangzi (Chang) valley and millet in northern China date from the early seventh millennium BCE. These locations were also where pigs, chickens, and, later on near the Yangzi, water buffalo were domesticated. Other crops were domesticated around 3500–2500 BCE in central Mexico (maize, amaranth, and squash), the central Andes (quinoa), and eastern North America (marsh elder, squash, goosefoot, and sunflower). In several other places, independent transitions to agriculture occurred well before the Christian era. Such locations include the Amazon basin (manioc, peanut), the Sudan region of Africa (millets, sorghum), the West African forest (yams), and Southeast Asia–New Guinea (taro, yams, banana, and sugarcane).

If agriculture arose in several locations from foraging, a logical question is why it did not happen earlier. The species that were domesticated had survived the Pleistocene epoch, though because of prevailing cold and aridity, many occupied smaller ranges than in the Holocene. Some experts argue that Holocene climatic stability was essential for agriculture, whereas others suggest other reasons, including prior global population growth, increased sedentism (permanence of settlement), and social change.

By 1500 CE, agriculture had spread across most of the world. Australia, the world's driest inhabited continent, was nonagricultural, although the Aborigines cultivated some plants. During the colonial era, agriculture reached nearly all the remaining areas except where it was not practicable.

John R. and William H. McNeill, in their book *The Human Web*, and Ian Morris, in *Why the West Rules . . . For Now*, use the domestication of plants and animals as an example of evolutionary change of human survivability (measured as the greater share of energy flows on the planet going to one species: humans).

Agricultural Intensification

Intensification is doing more to get more. Inputs, such as weeding and fertilizing, as well as the use of herbicides and insecticides, bring quick returns but must be repeated for continuing benefit. Increased demand drives intensification. Population growth is historically the greatest source of increased demand. Prestige production, meaning growing food for consumption at feasts and for display, has often been a factor.

Agricultural intensity may in turn affect population through improved nutrition, earlier marriage, and incentives to migrate. Increasing one agricultural input stimulates further intensification, for both economic and ecological reasons. Capital improvements encourage further inputs to maximize returns and repay the investment.

Growing populations have fed themselves through both intensification and expansion of agriculture. In some places, cultivation began in mostly *extensive* forms, that is, comparatively less intensive ones, often in which long fallow periods alternated with one or two seasons of cultivation. The fallow periods helped maintain soil fertility without the use of fertilizers. Elsewhere, cultivation began in at least moderately intensive forms, in small, suitable patches, and later spread.

Zonal and Regional Comparisons

Agriculture evolved in distinct ways in different climatic zones. Farming in arid and semiarid climates evolved around the need for moisture, including Mediterranean climates, with their mild and wet winters and hot and dry summers. Methods were often parallel in separate places. (Roughly half of the world's "Mediterranean" climate regions are found near the sea of that name; other so-called Mediterranean regions are found in Australia, South Africa, Chile, and the west coast of the United States.)

Moisture-conserving, or dry farming, techniques include spacing plants to reduce transpiration (i.e., water leaving the plant) and mulching with crop residues, volcanic ash, and other materials to reduce evaporation. Where draft animals are available, land is sometimes fallowed for a year; shallow tillage creates a layer of dust that conserves moisture for the next crop. Farmers have used animal manures, along with dry farming methods, in the Mediterranean since the classical era and have long exploited high groundwater areas and seepage zones. Gardens were excavated to a level where roots reached moist soil just above the water table in a few places in the Arabian and Sahara deserts and in Peru's coastal desert, where they were prevalent when the Spanish arrived.

Throughout arid and semiarid regions of Africa, Southwest Asia, and the Americas, early farmers exploited floodwaters. In *décrue* cultivation, crops are planted as slowly receding floodwaters uncover moist soils. (*Crue* is French for "flood.") In the right situations, yields are excellent and sustainable with little effort. The deep alluvial soils are often highly fertile, and each year's deposit of silt provides an increment of plant nutrients. Another strategy is to exploit runoff from brief rains by some means of diversion.

True irrigation can be simple and small-scale, or it can entail prodigious engineering works to move water long distances. Irrigation may involve continuous diversion from rivers and lakes or may lift water from them or from wells. Among the most awe-inspiring examples were fourteenth- and fifteenth-century Incan systems of tunnels, aqueducts, and canals spanning the high Andes. The Sabeans living in Marib (modern-day Yemen) built a large agricultural diversion dam in roughly 500 BCE. In North America, by 200 BCE the Hohokam tribes located in current-day Arizona had established extensive irrigation canals drawing from the Gila and Salt rivers. The fate of the Hohokam is uncertain: they vanished from the archaeological record around the fifteenth century. Some believe their water supply may have become oversalinized, while others believe it had to do with the arrival of the Spanish conquistadors.

To make use of scant rainfall, farmers from North Africa into parts of Central Asia planted around the bases of sand dunes, which trap rainfall. Around the Mediterranean, grapevines, olives, and other deep-rooted perennials tap receding soil moisture during the long, dry summers. Ancient Egyptian agriculture used a large-scale elaboration of *décrue* cultivation, and, before 3000 BCE, began building levees across the Nile floodplain.

An important climatic factor is the amount and duration of the annual rains. In the tropics, where frequent rains last nearly all year, excess water percolating through the soil leaches nutrients. Dense forests suggest fertility, but most of the nutrients are retained in the biomass. Dead vegetation quickly decomposes, releasing nutrients that root systems take up. As rainfall decreases, the natural cover becomes open forest, bush or thorn scrub, and grassland or savanna, containing progressively fewer nutrients in its biomass, but usually more in the soils, which are not so highly leached.

Systems incorporating long fallow periods are common in the tropics. A widespread form of extensive cultivation is known as "slash-and-burn." Cultivators cut undergrowth, then trees, and burn the cut vegetation once it dries. The first crop benefits from nutrients in the ashes and decaying slash. After one to three crops, the cultivators abandon the plot before invading weeds and lowered yields make the effort unrewarding. A fallow period of five years to several decades passes before a new cycle begins.

Permanent cultivation systems are varied. For some, adequate yields may continue for many years, even without soil amendments. Tree crops, alone or with interspersed annuals, give more sustainable yields, possibly because their deep root systems efficiently cycle nutrients. In most locations, however, something must be done to maintain fertility.

Cultivators use animal manures and, in a few places (mostly in Southeast Asia), use what is delicately called "night soil." Vegetation from bordering areas, along with weeds and residues from the previous crop, may be dug into mounds or ridges where the crop grows. The African savannas supported systems that combined grazing with manure-supported cultivation.

"Wet" systems like Asian paddy fields take advantage of nutrients available in the accompanying water or by alternating waterlogging and drying. A technique fairly widespread in the precolonial Americas, and also reported in parts of New Guinea and Africa, involved raising soil beds higher with the help of mud and organic sediments from surrounding ditches.

In places with humid, temperate climates, such as East Asia, eastern North America, and Europe, agricultural evolution diverged early on, despite some climatic similarities. A crucial difference lay in the use of grazing animals in agricultural systems.

Precolonial eastern North America lacked domesticated grazing animals, and foraging continued to be important long after native plants were domesticated. Much of the cultivation appears to have been in small, easily cultivated plots in river floodplains. Maize was introduced from Mexico before the Christian era, but was unimportant until circa 800 CE. From then until 1500, rapid population growth and agricultural expansion took place from the Mississippi valley to the Atlantic seaboard. Colonial-era descriptions are mostly of slash-and-burn cultivation, but archaeology has revealed signs of tillage suggesting earlier, more intensive systems, particularly in the Mississippi valley, before introduced diseases caused massive and widespread depopulation.

Agriculture spread rapidly on the Asian mainland from the Yangzi valley and the northern interior of China. Rice cultivation reached Japan around 300–400 BCE, although there was some earlier cultivation of native plants. In China, the Korean Peninsula, and Japan, grazing animals—oxen, water buffalo, and horses—were mostly used for draft and transport (especially horses). Pigs and poultry, which graze little, provided the most meat, and dairying was rare. When iron-shod plows became available in China, during the Han dynasty (206 BCE–220 CE), the use of draft animals increased.

Writers of the Han dynasty described the use of fertilizers and the planting of one or more crops each year, although these methods were not yet universal. Intensive methods emerged throughout much of the region in the second millennium CE, as average farm sizes shrank further. In dry areas of China and in certain hilly areas of China and Japan, slash-and-burn methods persisted into the twentieth century.

Fields replaced forest over succeeding millennia. Plowing was known by the late fourth millennium BCE or earlier, and evidence of manure use is almost as old. Wet areas, or meadows, were typically used to produce hay for winter fodder. During periods of chaos or plague, depopulation led to agricultural contraction.

Industrial Agriculture

Industrial agriculture is supported by an industrial system that supplies machinery to save labor, chemical fertilizers to raise yields, and a variety of chemicals to control weeds, pests, and diseases. Like modern industry, industrial agriculture uses large amounts of energy, largely substituting that for human and animal effort. It has also come to rely on modern scientific research, although applications of science and of industrial inputs are not always dependent upon one another.

The 1830s marked the first substantial use of industrial machinery and inputs in agriculture. Horse- or oxen-pulled reapers were first marketed in Scotland and the United States, and the inexpensive iron of the Industrial Revolution made them (as well as later machines) affordable. Chemical fertilizers in early use in England included sodium nitrate from natural deposits in Chile and ammonium sulfate, a by-product of coal gas and coke manufacture.

For decades, Western Europe adopted chemical fertilizers, while mechanization was more prevalent in North America (the high prices of land in the former and of labor in the latter influenced these trends). Steam tractors were used extensively in the nineteenth century, but gasoline tractors competed seriously with draft animals after 1910 in North America, Australia, and Great Britain, and in the 1930s in the Soviet Union. In Japan and parts of Europe, small farm size meant that the adoption of tractors lagged behind the use of chemical fertilizers between the world wars. In less developed regions, both trends began after World War II and are still far from universal.

The impact of pesticides on the cultivation of major field crops was minor until after World War II. The insecticide lead arsenate was in widespread use after 1867, as were several fungicides for fruit by the late nineteenth century. During the 1950s and 1960s, farmers widely adopted DDT and organophosphate insecticides, developed during World War II, and in many developing nations their adoption outpaced that of other industrial inputs.

Mechanization increased labor efficiency. A team of oxen needs weeks to plow what a large tractor can plow in a single day. Mobile machines exist for most basic tasks of cultivation, except that most weed suppression is now accomplished with chemical herbicides. Although the United States is a net exporter of farm commodities, its farms are mechanized, and less than 2 percent of the US workforce is employed on farms or in direct support. In most historic preindustrial economies, agriculture employed at least 70 percent of the workforce.

Industrial agriculture has boosted yields, but has had long-term costs to the environment. Force pumps, which tap water deep underground, have made possible a large extension of irrigated farmland. Chemical control of diseases and pests has reduced losses, at least in the short-term. The increased yields obtained through the use of chemical fertilizers are also obtainable with manure and other "amendments" long used in preindustrial agriculture. Chemical fertilizers increase the supply of amendments, however, allowing high application rates to be more common.

The plant nutrient that is most often limiting is nitrogen. Biological nitrogen fixation, by which microorganisms in the soil and in association with the roots of legumes, such as beans, makes atmospheric nitrogen available to crops, is crucial to preindustrial agricultural systems. Modern scientific breeding has created high-yielding crop varieties that respond to levels of available nitrogen that biological nitrogen fixation cannot provide.

Preindustrial and Industrial Agriculture Compared

Agriculture inevitably alters natural ecosystems and has environmental impacts. Even grazing on natural grasslands favors some plant and animal populations over others. Conversion of wild areas to farmland reduces the habitat of many species. The erosion of soil by wind and water has degraded or destroyed vast areas that grew food several millennia ago.

Some modern trends increase environmental impact, but how much of this is avoidable? Specialization in a limited number of varieties of many major crops threatens genetic diversity. Seed banks have been proposed and created (the best-known of which is the Svalbard Global Seed Vault, on the arctic Norwegian island of Spitsbergen), although some dispute their adequacy. In some places, the use of tractors has increased tillage and accelerated erosion. Reduced tillage methods, often substituting chemical for mechanical weed control, is one remedy. In some temperate regions, the heavy use of chemical fertilizers has replaced the growing of cover crops that once provided winter ground cover. A partial remedy is to leave crop residues on the ground all winter long.

Pollution from agriculture is increasing as excess nutrients from agricultural fields contaminate water systems. The concentration of livestock has turned manure from a valuable fertilizer to a pollutant when it is not contained. Direct threats to the health of humans and ecosystems led to the ban on DDT in the 1960s in most of the developed world and on other insecticides that persist in the environment. Pesticides are either persistent and nonpersistent, depending on how long they stay in the environment; some pesticides marketed as being environmentally friendly may in fact belong on the persistent pesticides list.

When a developed nation bans a chemical like DDT, that does not mean the rest of the world stops using it. US pesticide manufacturers continued to export DDT to the developing world after it was banned in the United States. Pesticides sometimes come back into the developed world as residue on agricultural produce, in what is often called the "circle of poison." Some shorter-lived but toxic pesticides and herbicides remain in use. Pests and some weeds resist them, and they have destroyed beneficial organisms that help control pests. "Integrated pest management" and "integrated weed management" use chemicals selectively and with restraint. In addition, natural pest killers such as bats or ladybugs, which eat aphids, are used whenever possible.

Without the benefits of industrialized agriculture, it would not be possible to have 7.1 billion humans on the planet. Industrialized agriculture, however, depends upon nonrenewable resources. Fossil fuels power machinery and enter into every major industrial input, adding to global warming, although agriculture directly accounts for only about 5 to 10 percent of fossil fuel use in industrialized countries. Among the elements in chemical fertilizers, only nitrogen is renewable, because it naturally cycles among air, crops, and soil (other elements such as potassium and phosphorus come from mineral deposits).

Biological Exchanges

For millions of years, most terrestrial species "stayed home." Geographic barriers, such as oceans and mountain chains, inhibited migrations and divided the Earth. Only birds, bats, flying insects, and good swimmers consistently bucked the trend. A few other species did so occasionally, thanks to sea-level changes and land bridges or to chance voyages on driftwood and other vegetation. Natural evolution took place, for most species, in separate biogeographical provinces.

Intracontinental Biological Exchange

This long phase in the history of life ended when human beings began their long-distance migrations. Deep in prehistory, people—or hominids at any rate—walked throughout Africa and Eurasia, occasionally bringing a plant, seed, insect, rodent, or microbe to a place it would not have reached on its own. With plant and animal domestication some 10,000–12,000 years ago, people began to do this on purpose and more frequently.

Most of the plants and animals susceptible to domestication were found in Eurasia, and the east-west axis of that continent eased the spread of those plants and animals sensitive to climate conditions or to day length (several flowering plants take their cues to bloom from day length). The author Jared Diamond has pointed out that because of the relative sameness of daylight and climate along Eurasia's long east-west axis, agricultural technologies have been able to pass more easily across this vast region over the millennia.

The spread of domesticated plants and animals took a few millennia, a disruptive process in which species invaded local biogeographic provinces. It also proved disruptive historically, obliterating peoples who did not adapt to the changing biogeography, the changing strains of disease, and the changed political situations brought on by the spread of farmers, herders, and eventually of states. Out of this turmoil of Afro-Eurasian biological exchange emerged the great ancient civilizations, from China to the Mediterranean. They all based their societies on intersecting but not identical sets of plants and animals.

Biological homogenization within Afro-Eurasia had its limits. The links between North Africa, say, and East Asia before 500 BCE were slender. Varying topography and climate also checked the spread of species. The process presumably accelerated when large states created favorable conditions for the movement of goods and people. The era of the Han and Roman empires, for example, when merchants transversed the Asian Silk Roads, unleashed a small flood of biological exchanges.

Two other periods in Eurasian history heightened biological exchange. The first of these occurred during the early Tang dynasty in China, in the seventh and eighth centuries. The Tang rulers showed keen interest in all things foreign, importing exotic creatures, aromatic plants, and ornamental flowers. The Tang were culturally receptive to strange plants and animals, but there was more to it than that. Their political power on the western frontier promoted the trade, travel, and transport that make biological exchange likely. A handful of large empires also held sway throughout Central Asia, making the connections between China, India, Persia, and southwest Asia safer than usual. This geopolitical arrangement fell apart after 751 CE, when Muslims defeated Tang armies in Central Asia, and after 755 CE, when a rebellion attempt shook the Tang to its foundations and millions lost their lives. Thereafter, both the stability of the geopolitical situation and the receptivity of the Tang to anything foreign waned, and the opportunities for biological exchange grew scarcer.

The other moment of heightened biological exchange within Eurasia came with the Pax Mongolica of the thirteenth and fourteenth centuries, a period of relative stability after the victories of the Mongols in Europe and Central Asia. By this time, most of the feasible exchanges of plants and animals had already taken place. The heightened transport across the desert-steppe corridor of Central Asia may have brought carrots and a species of lemon to China, and a form of millet to Persia. Quite possibly, it also allowed the diffusion from Central Asia of the bacillus that causes bubonic plague, provoking the famous Black Death, the worst epidemic in the recorded history of western Eurasia and North Africa.

Although this process of Eurasian (and North African) biological exchange never truly came to an end, it slowed whenever political conditions weakened interregional contacts. It also slowed in general after around 200 CE, with the erosion of the Pax Romana and Pax Sinica, which had encouraged long-distance travel and trade within Eurasia. By that time, sugarcane had taken root in India, spreading from its New Guinea home. Wheat had spread widely throughout most of its potential range, as had cattle, pigs, horses, sheep, and goats.

Meanwhile, on other continents, similar if smaller-scale processes of biological exchange and homogenization were in process. In the Americas, maize spread from its Mesoamerican home both north and south. In Africa, the migrations of Bantu speakers 2,000 years ago probably diffused several crops throughout eastern and southern Africa, and possibly brought infectious diseases that ravaged the indigenous, previously isolated, populations of southern Africa. These events in Africa and the Americas, too, must have been biologically and politically tumultuous, although the evidence is sparse.

In biological terms, biological exchange selected for certain kinds of species that coexisted easily with human activity, such as domesticates, commensals (species such as rats that obtain food from another species, without being either beneficial or harmful to that species), and plants that thrive on disturbed ground, most of which we call weeds.

Intercontinental Exchanges and Invasions before 1400 CE

Intercontinental biological exchange has a long pedigree. The first people to migrate to Australia may have accidentally brought some species with them 40,000 years ago. About 3,500 years ago, later migrants to Australia purposely brought the dingo (a large dog), the first domesticate in Australian history. The dingo quickly spread to all Aboriginal groups outside of the island of Tasmania, and also formed feral packs. As a hunting dog, it was so effective that it led to the extinction of some indigenous mammals. The dog was also the first domesticated animal in the Americas, brought across the Siberian-Alaskan land bridge with some of the first settlers during the last Ice Age. Dogs probably played a significant role in reducing the populations of large mammals, many of which became extinct around the time humans arrived in North and South America.

Initial human settlement of unpopulated islands also wrought major ecological changes throughout the southwest Pacific and Polynesia, including numerous extinctions, from about 4,000 years ago through the colonization of New Zealand roughly a millennium ago. All these instances were invasions of lands that had no prior exposure to humanity and its fellow travelers, or to the intensified fire regime that human presence normally brought. This explains the dramatic effects, particularly the rash of extinctions, that followed upon human settlement of Australia, New Zealand, and the Americas.

People began to transport animals, plants, and pathogens from one human community to another across the seas. In many cases, the only evidence for such transfers is the existence of the imported species. The sweet potato, a native of South America, somehow arrived in central Polynesia by 1000 CE and spread throughout Oceania. A second mysterious transoceanic crop transfer took place across the Indian Ocean some time before 500 CE. Bananas, Asian yams, and taro somehow arrived in East Africa. These crops had much to recommend them, because they do well in moist conditions, whereas the millets and sorghum that Bantu expansion brought into central and southeastern Africa were adapted to dry conditions. Bananas, taro, and yams were probably introduced to East Africa more than once, and almost surely were brought again in the Austronesian settlement of Madagascar that took place just before 500 CE. These Asian crops assisted in the epic (but unrecorded) colonization of Central Africa's moist tropical forests by farmers, as well as in the settlement of Madagascar.

Several other significant intercontinental biological transfers took place before 1400, mainly between Africa and Asia, a route that posed minimal obstacles to sailors. Africa's pearl millet, derived from a West African savanna grass, is the world's sixth most important cereal today. Introduced into India some 3,000 years ago, it accounts today for about 10 percent of India's cereal crop coverage. East African sorghum entered India at about the same time, and eventually became India's second most important grain after rice. Finger millet, also from Africa, made it to India only around 1,000 years ago. The transfer of African crops to South Asia provided India with drought-resistant dryland crops, opening new areas to settlement and providing a more reliable harvest where water supplies were uncertain. These examples suggest a world of crop exchange—and probably weeds, diseases, and animals too—around the Indian Ocean rim from around 3,000 to 1,000 years ago.

While South Asia received new crops from Africa, it sent others to the Middle East and the Mediterranean. Between the tenth and thirteenth centuries, Arab trading networks, facilitated by the relative peace facilitated by the Abbasid caliphate (750–1258), brought sugar, cotton, rice, and citrus fruits from India to Egypt and the Mediterranean. These plants, and the cultivation techniques that came with them, worked a small revolution on the hot and often malarial coastlands of North Africa, Anatolia (modern Turkey), and southern Europe. They caused many coastal plains to be brought under cultivation on a regular basis, often for the first time since the Roman Empire.

A second avenue of exchange involving the Mediterranean basin linked it to West Africa. Although this was not genuinely an intercontinental exchange, the Sahara functioned like a sea for several millennia, as the use of the Arabic term for shore (Sahel) for the West African desert edge implies. A thousand years before Columbus crossed the Atlantic, some unknown soul crossed the Sahara, reuniting the Mediterranean and the Sahel, which the increasingly arid Sahara had divided since about 3000 BCE. Trans-Saharan trade developed in salt, slaves, and gold. Large horses made their debut in West Africa via trans-Saharan trade, which took the social, economic, and political history of West Africa in a new direction.

Long before the great age of oceanic navigation, the links of trade and colonization in the Pacific, in the Indian Ocean, and across the Sahara brought biological exchanges that had a powerful influence on the course of history. The further exchanges following the voyages of Columbus, Magellan, Cook, and others extended this process to lands once quite separate in biological (as well as other) terms.

Biological Globalization after 1400

After 1400, mariners linked almost every nook and cranny of the humanly habitable Earth into a biologically interactive unit. It became a world without biological borders, as plants, animals, and diseases migrated wherever ecological conditions permitted their spread, although how soon and how thoroughly they did so often depended on patterns of trade, production, and politics.

Columbus inaugurated regular exchanges across the Atlantic, whereby the Americas acquired a large suite of new plants and animals, as well as diseases that depopulated the American hemisphere between 1500 and 1650. Simultaneously, Africa and Eurasia acquired some useful crops from the Americas, most notably potatoes, maize, and cassava (or manioc). Ecosystems and societies in the Americas were remade, with new biologies and new cultures, as they were in Africa and Eurasia, although less catastrophically.

The new food crops fueled population growth in Europe and China, and possibly in Africa too, although there is no firm evidence of this. Maize and potatoes revolutionized agriculture in Europe, as maize and sweet potatoes did in China, by allowing more intensive production and new lands not suited to wheat, barley, rye, or rice to come into production. In Africa, maize, cassava, and peanuts became important crops. Some 200 million Africans today rely on cassava as their staple food. Many of the rest, mainly in the south and east, rely on maize.

These modern exchanges had political meanings and contexts. European imperialism, in the Americas, Australia, and New Zealand, promoted and was promoted by the spread of Eurasian animals, plants, and diseases. Europeans brought a biota that favored the spread of European settlers, European power, and Eurasian species, and created what Alfred Crosby, the foremost historian of these processes, called "neo-Europes"—including Australia, New Zealand, most of North America, southern Brazil, Uruguay, and Argentina.

Beyond the neo-Europes, there emerged in the Americas a neo-Africa. More than 10 million Africans arrived in the Americas in slave ships. Yellow fever and malaria, which profoundly influenced settlement patterns in the Americas, arrived on American shores in those ships. They also brought West African rice, which became the foundation of the coastal economy in South Carolina and Georgia in the eighteenth century, and was important in Suriname as well. Other African crops came too: okra, sesame, and coffee. African biological impact on the Americas continued beyond the end of slave trading. African honeybees imported into Brazil crossbred to create an "Africanized" bee that since the 1950s has colonized much of the Americas.

Sailing ships did not prove hospitable carriers to every form of life. The age of steam, and then of air travel, broke down further barriers to biological exchange, adding new creatures to the roster of alien intruders and accelerating the dispersal of old and new migratory species alike. The advent of iron ships toward the end of the nineteenth century, for example, opened a new era in biological exchange involving species living in the world's harbors and estuaries. After the 1880s, iron ships began to carry water as ballast. Soon special water ballast tanks became standard, and so, for example, a ship from Yoko-hama bound for Vancouver would scoop up a tankful of water and a few marine species from Japanese shores, carry it across the wide Pacific, then release it in Puget Sound before taking on cargo.

In the 1930s, Japanese clams hitched such a ride and upon arrival began to colonize the seabeds of Puget Sound, creating a multimillion-dollar clam fish-ery in British Columbia and Washington State. A jellyfish that devastated Black Sea fisheries arrived from the East Coast of the United States in about 1980. The zebra mussel, a native of the Black and Caspian seas, colonized the North American Great Lakes and associated river systems from a beachhead estab-lished near Detroit in 1985 or 1986 and is also a problem in much of Europe, including the British Isles. A more recent invader of the North American Great Lakes is a crustacean called the fishhook flea, also a native of Caspian and Black Sea waters. It first appeared in Lake Ontario in 1998, most likely introduced from ballast water brought by ships from the Baltic Sea, and is now in all the Great Lakes and New York's Finger Lakes, menacing sport and commercial fisheries and disrupting the lakes' food web.

The technology of transportation influences the process of biological exchange. The invention of ships and their ballast tanks, of railroads and air-planes, all led to changes and surges in the pattern of biological exchange. This provides one rhythm; another is political.

Some states and societies imported exotic species. Monarchs of ancient Egypt and Mesopotamia maintained gardens and zoos filled with exotic plants and animals. Thomas Jefferson (1743–1826), the third president of the United States, tried his best to establish rice, wine grapes, and silkworms in Vir-ginia. Later, the US government employed plant prospectors who scoured the globe for useful species, and brought tens of thousands back. Today the United States, Australia, New Zealand, and many other countries spend vast sums to prevent the importation of unwanted species, hoping to forestall biological invasions.

The changing nature of geopolitics has also affected biological exchange. Trade and travel—and presumably biological exchange—flourished in peace-time and contracted in times of war and piracy. Probably eras of imperial power, when a single power enforced a general peace, provided the best polit-ical environment for biological exchange. Furthermore, imperialism inspired and eased the process of collection over vast distances. Kew Gardens, a botan-ical garden outside of London, proved a crucial link in transferring rubber seeds from Brazil to Malaya at the end of the nineteenth century, which started a new plantation economy in Southeast Asia.

The spread of pathogens that cause disease and epidemics, a key element of biological exchange that until recently has drawn little attention, is a topic we turn to in the following section.

Human Disease

The environments of human diseases have geographical/climatic, cultural, and biological factors. When humans lived in smaller, more mobile communities, they were not disease free, but did not suffer the epidemic diseases and crowd-type diseases that came later in history. The infectious diseases that contributed to great human morbidity and mortality likely originated mostly in southeastern Asia and Africa. In the Old World, several environmental factors combined to change the pattern of disease. The development of agriculture involved larger human communities (discussed in the next section, on urbanization), which were more stationary and had less dietary variety. Although agriculture offered control over the food supply, weather-related events made it more vulnerable.

Famine and malnutrition lowered human resistance to pathogens (agents of disease). Irrigation techniques created breeding grounds for disease organisms, and associated plowing techniques may have increased the risk of fungal diseases. The semipermanent settlements associated with agriculture increased the likelihood of contaminated water supply and pathogen exposure due to human contact with compromised environments.

Any discussion of the role of environment in the development of human disease must consider both cultural and geographical factors. Climatic factors such as humidity, rainfall, and temperature can affect the nature and impact of human disease. Geographical factors can influence human settlements, population movements, warfare, trade, and other aspects of culture that play key roles in human disease. Culture itself—in particular, the dynamics of food procurement, settlement patterns, sanitation practices, trade, and warfare—influences the occurrence and impact of human disease. These factors are linked in complex and dynamic ways with a great variety of expression. Human disease needs to be viewed in its environmental perspective.

Life forms transferred to humans by various modes cause disease; their impact can be worsened by social factors. Starvation and malnutrition weaken the human host and lead to failure of the immune system. In some cultures, poverty places humans in proximity to nonhuman animals and their potential disease vectors. Irrigation provides breeding grounds for mosquitoes.

Early Cities

Some historians contend that the world's first cities emerged as consequences of the formation of the state. Others hold that early states emerged from cities. Whichever the sequence, there were several areas where cities first appeared: lower Mesopotamia in southwestern Asia, where Sumerian cities appeared 5,000 years ago; the Indus River valley of Asia, 4,500 years ago; the northern China plain, 3,000 years ago; Mesoamerica, 2,500 years ago; the central Andes and Peruvian coast, 2,000 years ago (although recent evidence places the age of the pyramids at Caral in Peru at 4,600 years); and the Yoruba territories in western Africa, 500 years ago (and beginning at the same time in Zimbabwe and the lower Congo River valley).

As these urbanized societies slowly spread their technologies and institutions, cities began to appear in a number of other areas: Japan, the Indian Deccan region, southwestern Asia, the Mediterranean, and Europe. Beyond these areas, in the New World in particular, urbanization came as a consequence of colonial expansion and imposition, often with explicit planning guidelines, as in Spain's sixteenth-century Law of the Indies, issued to rule American and Philippine possessions of the crown.

Cultural evolution in each of the zones of primary urban generation characteristically began with domestication of plants and animals, and with the emergence of class-based societies. The formation of military and religious elites followed, and they gathered clans into states and used their power to extract surpluses from village farmers. In such states, hierarchies of specialized institutions developed and exercised authority over territory and maintained order within their populations. Within the settlements resided specialists who began as temple and palace functionaries and later evolved into producers for the market. Similarly, the merchants who conducted long-distance trade evolved from the networks of tribute that military action earlier had secured. Astronomy was important in the regulation of the rhythms of agriculture, as well as in the physical plans of the capital cities, which included astronomical orientations and cardinal (usually north-south and east-west) street alignments.

Most classical capital cities were small and compact. Levels of urbanization never exceeded 10 percent, including any secondary centers that developed, and these were few and small. As late as 1700, there were probably no more than fourteen cities in the world with populations exceeding 200,000 found in China, Japan, the Mughal empire (which covered most of the Indian subcontinent), Persia, the Ottoman empire, and in Europe. Perhaps another fifty cities exceeded 50,000.

Despite their small size, capital cities served as the focal points of distinct city-centered "world economies," economically autonomous sections of the planet able to provide for most of their own needs. Each was surrounded by an immediate core region (called on to provide foodstuffs and migrants and within which modification of the Earth was greatest), a developed middle zone exploited for transportable resources and products, and a vast and untouched periphery.

Now we return to the discussion on the human community's increasingly productive and destructive use of the environment for subsistence, moving from agriculture and settlement to the age of industry.

❧ ❧ ❧

CHAPTER 3

The Shift to Cities and Industry

Examine the history of all nations and all centuries and you will
always find men subject to three codes: the code of nature,
the code of society, and the code of religion...these codes
were never in harmony.

—Denis Diderot (1713–1784)

For nearly all of their existence on earth, humans have used fairly simple technology, such as hand-held tools, animal-powered plows, irrigation channels, and fire, to alter their environment. In the eighteenth century, a major transformation began, one perhaps more significant than the earlier shift from foraging to farming. This included the Industrial Revolution, the rise of industrial agriculture, and the growth of cities.

The Industrial Revolution

During the Industrial Revolution, economies shifted away agriculture and the production of goods in homes to the production of goods in factories. This resulted in a corresponding growth of population centers and transportation networks for the distribution of those goods. The first Industrial Revolution occurred in England between 1740 and 1780. From the standpoint of the environment, the significance of the Industrial Revolution was the increased use of fossil-based energy sources and other natural resources, placing ever-increasing demands on the environment.

This period saw a fundamental shift from societies that had high birth and death rates to societies that had low birth and death rates, as they are in the developed world today. The Population Reference Bureau estimates that in

2011, the figures were 19 births and 8 deaths per 1,000 people, with 360,000 people being born each day and around 150,000 people dying each day.

Preconditions for the Industrial Revolution

The Industrial Revolution could not have occurred until the Agricultural Revolution that preceded it had a chance to get under way in England, making it possible to produce increased amounts of food using fewer workers. Industrialization affected agriculture and the economy. Inventions such as the spinning jenny and the cotton gin increased the profits gained by planting cotton. The improvements in steam engines allowed operators to pump water out of coal mines, dramatically increasing access to an important fossil fuel resource. The mechanical reaper increased the speed with which crops could be harvested. The development of machine tools made the manufacture of farm implements easier. Canals and railroads facilitated the transport of food into urban areas and brought manufactured goods to local markets.

Industrialization required a stable currency and banking system. Industrial workers needed to be confident that the coins or banknotes with which they would be paid would be accepted in exchange for goods and services. Industrialists often needed to borrow the funds to build and equip factories. Banks and other lenders needed to be sure that the currency in which they would be repaid would have the same buying power as it did at the time it was borrowed.

Other factors that contributed to the Industrial Revolution include the earlier scientific revolution, associated with English physicist and mathematician Sir Isaac Newton, and the religious thought of the Protestant Reformation, which established a link between material prosperity and divine favor.

New Sources of Energy and Industries

A case can be made that the precipitating event of the Industrial Revolution was a change in the ecology of England—the depletion of English forestland that resulted from burning wood for warmth and cooking and for use in construction and in such activities as glassmaking. Coal, the first fossil fuel to be widely used, was an even more efficient source of heat than was wood, and iron would turn out to be a far more durable building material. Coal and iron had to be mined, however, and mines tended to fill up with water. A primitive steam engine, known as the "miner's friend," was patented in 1698 and used to remove water from mine shafts. This technology was improved upon, with Scottish inventor James Watt (1736–1819) finally introducing an efficient engine, and by 1800 more than five hundred steam engines were in commercial use.

The steam engine represented the first use of heat energy to perform mechanical work. The only power sources available to industry had been human and animal muscle, and the energy of flowing water and the wind. A steam engine, however, could be installed anywhere water and fuel were available.

The basic operations in the production of cloth are spinning, which winds short fibers into thread or yarn, and weaving to form cloth. The first major invention in textiles during the Industrial Revolution was the flying shuttle loom in 1773, followed by the spinning jenny in 1764 and then an improved spinning machine. By 1778, twenty thousand spinning machines were in operation. Steam engines were installed as "prime movers" in the textile factories, providing mechanical energy to the spinning machines and, later, power looms. In 1793, the American inventor Eli Whitney (1765–1825) began production of the cotton gin, a device that speeded the removal of cotton seeds from cotton fiber, assuring an increased supply of raw cotton. (With his invention, Whitney inadvertently helped to extend the reign of slavery for many decades. The cotton gin radically reduced the price of producing cotton and thus made the southern United States a primary supplier of slave-picked cotton for textile mills in Great Britain, as well as the northern states.)

The extraction of iron from its ores and its use in construction greatly expanded during the Industrial Revolution. (Iron smelting occurred much earlier in China, which leads to the interesting question of why an industrial revolution did not take place there.) Iron production is costly in energy terms. Wood does not burn at a high enough temperature to be used directly to heat the ore, so charcoal had been used instead. The copper industry had begun using coke, a purified form of coal, as a more reliable heat source. Iron and copper smelting both remained small and expensive operations until coal mining and transportation became more industrialized and efficient.

Numerous advances in land and water transportation marked the Industrial Revolution. In 1760, the British Parliament approved using public funds for the construction of the Bridgewater Canal to carry coal from the mines at Worsely to the manufacturing city of Manchester, which reduced the cost of coal to the textile manufacturers by about 50 percent. There followed a period of extensive canal building, with nearly fifty additional canals being built by the end of the century and industrialists establishing factories on sites that would reduce their transportation costs.

Britain enjoyed a strong navy and a merchant marine fleet. The invention of a chronometer, which measured time onboard a ship and thus could be used, together with astronomical observations, to determine longitude, made navigation on the open sea much safer. Sailing remained dependent on wind, however, and thus somewhat unpredictable until the American inventor Robert Fulton invented a practical steamboat in 1787. The steam locomotive appeared in 1802. The railways would become the biggest consumer of industrial iron, required both for the steam-powered trains and the rails they rode on.

Nations in the developing world envy the prosperity of the United States, Britain, and the countries of northern Europe as industrialized nations, and many poorer nations have set upon an agenda of industrialization. Becoming an industrial power, however, is difficult without a stable currency and government and a literate population. Cultural considerations also come into play; not all cultures share the European notion of progress, and many continue to deal with the lasting legacies of colonialism. Industrialization has expanded in the Pacific Rim as multinational corporations find it easier to manufacture electronic goods in countries with cheaper labor and less stringent environmental laws. How best to facilitate the industrialization process without further deterioration of the global environment remains an issue of international debate.

Industrialization Spawns Urbanization

By 1800, new forces were at work that would redraw the world map of urbanization: what demographers refer to as "rural push" and "urban pull." Urban growth in England was already accelerating outside London. The precipitating factor was the initial wave of the Industrial Revolution, brought about by major advances in the cotton and iron industries, the first flush of factory building, and significant improvements in waterborne transportation as canals were built and rivers were dredged. The new urban centers were either mill towns in which the workers resided within walking distance of the factory, cities such as Birmingham that had at their cores specialized industrial districts, or centers of control and finance of the new economy, such as Manchester. Rural-to-urban migration fed the demand for labor and new factory production displaced workers from occupations such as hand spinning and hand-loom weaving.

Nineteenth-century Industrial Urbanization

Building upon this break with the past, accelerating technological change brought a new kind of city between 1800 and 1900, built on productive power, massed population, and industrial technology. By the end of the century, this new kind of city created a system of social life founded on entirely new principles, not only in Britain, but also in Europe, in the United States, and in pockets elsewhere across the globe. By 1900, the level of urbanization had reached 80 percent in Britain, exceeded 60 percent in the Netherlands and industrializing Germany, 50 percent in the United States, and 45 percent in France. Sixteen cities in the world exceeded 1 million in population, and 287 exceeded 100,000. Two great urban-industrial core regions shaped the world economy—western Europe and the northeastern United States, with a third emergent in Japan.

Cities grew as dense concentrations around their central business districts, which housed the headquarters of their corporations and centers of finance, together with agglomerations of downtown business, in locations that offered accessibility to the surrounding population. Urbanism became a new way of life as the increasing size and density of cities and the increasing heterogeneity of their immigrant populations produced social consequences such as individual freedoms and opportunities for social and economic advance. They also produced the inequality and alienation that could lead to unrest and revolution. The new cities also changed their local environments through the unrestricted discharge of effluents (into the air, water, and land and on to the hinterlands) from the efficient transformation of their resources into manufactured goods.

Twentieth-century Urban Growth

During the twentieth century, the urbanization level in advanced nations leveled off at around 80 percent, while technological change brought accelerating urban growth to most other parts of the world. By 2011, there were twenty-three "megacities" (as defined by the United Nations) exceeding 10 million people. The number of megacities is expected to rise to thirty-seven by 2025.

The new technologies changed the patterns of urban growth. The concentrated industrial metropolis had developed because centrality and proximity meant lower transportation and communication costs. Shortened distances also meant congestion, high rents, and loss of privacy. In contrast, the technological developments of the twentieth century made it possible for each generation to live farther apart and to rely upon distant information sources. Centralization and overall densities declined, producing far-flung metropolitan regions and emptying out the higher-density cores.

As these changes unfolded, researchers began to assess the modifications of the physical environment that this rapid urban growth and transformation produced. These modifications occurred at three geographic scales. First, locally, they altered the nature of the surface of the Earth, replacing the natural surface of soil, grass, and trees with urban surfaces of brick, concrete, glass, and metal at different levels above the ground. These artificial materials change the nature of the reflecting and radiating surfaces, especially the heat exchange near the surface. Second, regionally, they generated large amounts of heat and altered the atmosphere with emissions of gaseous and solid pollutants. At certain times of the year in midlatitude cities, artificial heat input into the atmosphere may approach or even exceed that derived indirectly from the sun. The heat island that results serves as a trap for pollutants. Third, globally, urban activities raise sulfur and carbon dioxide levels in the atmosphere and add to the greenhouse effect, to global warming, and to sea-level changes, which are likely to be of greatest consequence for major coastal cities.

One of the principal local effects of urbanization are land-use changes that affect the system of rivers and streams and the patterns of runoff. Runoff occurs more rapidly (because asphalt is not permeable) and with a greater peak flow under urban conditions. Water pollution changes the quality of downstream resources and the ecology of the riverine environment. The effects become pronounced downstream of larger cities, where natural flushing is incapable of preventing long-term damage.

Beyond such regional-scale consequences, urban activities produce carbon dioxide and fluorocarbons that in combination may affect future global climates and sea levels. These effects intensify as more than 50 percent of the world's more than 7 billion people have moved into cities. Many argue that urban areas have lower per capita impact than other geospatial arrangements. Others, including the sustainability pioneer William Clark, suggest that the sustainability of cities will be much more difficult to achieve than rural sustainability. Urban people today purchase rather than grow their own food, which comes from high-tech, productive farms, a sector of the economy called agribusiness.

Agribusiness

Agribusiness engages in producing, processing, and transporting agricultural products. It is what enables large numbers of people to live in cities. So defined, agribusiness involves three distinct systems and types of institutions with three distinct historical bases. The first is farmer-based agribusiness, a natural outgrowth of the operation of large numbers of small farms by owner-producers, usually family groups, for their own benefit. The second is elite agribusiness: government power limits access to farmland or farm resources to members of a social elite who, when engaged in farm production, use absentee ownership and unfree (e.g., serfdom, conscripted labor, slavery) labor. The third is industrial agribusiness, a consequence of industrialization and the accelerated process of societal economic specialization to which it has given rise.

Industrial agribusiness is now dominant in the production of fertilizers, herbicides and pesticides, farm equipment, and, increasingly, hybridized and genetically modified seed. In food processing and distribution, the longer and more complex the steps between the farmer and the final consumer, the more large-scale industrial agribusiness dominated.

Since the 1950s, the rapid growth of industrial agribusiness has been the trend worldwide. Although largely rooted in the United States, Australia, and the European Union, the effects have been global. Agribusiness has two major dimensions: the movement of work (other than immediate farm production) off the farms and the concentration of the business itself in fewer and larger firms.

Specialization and differentiation among agriculturally related occupations is the most comprehensive change. The proportion of the population directly engaged in farm production has decreased and total agricultural production has increased. Many related jobs are no longer rural and many rural jobs are no longer agricultural. The largest categories are agricultural processing and marketing, textiles, and agricultural wholesale and retail.

Global Agribusiness and the Environment

The rapid growth of industrial agribusiness raises the possibility that powerful firms such as Monsanto, DuPont, and Dow can use their resources to obtain the kind of exploitive control associated with elite agribusinesses.

Agricultural production will have to continue to increase, while becoming less reliant on traditional agribusiness inputs. The way to do this seems to be through biotechnology, which may take the form of genetically modified organisms (GMOs) or genetically engineered crops—the creation of modified crop varieties and other organisms by incorporating genetic material from other species in their genetic makeup.

Biotechnology also poses threats, however—one being further environmental pollution and threats to biodiversity (biological diversity as indicated by the number of species of plants and animals). Industrial agribusiness threatens biodiversity by the rapid mechanized destruction of the world's remaining rainforests to make room for industrialized farms' livestock-raising

operations. Another threat emerges from biotechnology: the use of "terminator" and transgenic genes that can alter the environment. ("Terminator" or
"suicide" seeds are seeds that are engineered in such a way that the second
generation of seeds is sterile, requiring the farmer to purchase more seeds
from the supplier. This is referred to as genetic use restriction technology, or
GURT.)

The worldwide Green Revolution stimulated the expansion of industrial
agribusiness. Governments, universities, and philanthropic foundations produced the core Green Revolution technologies on a not-for-profit basis. The
consequence has been that although food production has been able to outpace
population growth, environmental pollution caused by the chemical fertilizer,
herbicide, and pesticide use has accompanied it.

The Green Revolution

In the narrowest sense, the Green Revolution has adopted and spread specific
agricultural technologies that allow farmers to increase food production per
unit of land and per unit of labor. The Green Revolution is often divided into
a post-World War II first revolution that focused on increasing the ability to
produce more food by industrial means, and a second revolution, starting in
the 1980s, that focused on biotechnology and fine-tuning what crops would
grow best and where. Although corporate farms in developed countries have
adopted Green Revolution technologies on a large scale, their primary purpose
has been to serve the needs and interests of small-scale independent farmers
in underdeveloped countries. In this sense, the Green Revolution was not only
a technological or agricultural revolution but also a full-scale social revolution.

The Green Revolution has assured that for the present and the immediate future, world food production will exceed world food needs. Many countries formerly facing famine are now self-sufficient. At the same time, it has
changed the way the world's farmers relate to their social and technological
contexts. The Green Revolution came at a great cost to the environment, one of
which has been greater pest resistance to chemicals. This is similar to what has
happened with the overuse of antibiotics to treat sickness.

The core of the Green Revolution is a series of cultivars (domesticated
plant variety), mainly grain crops, called "high-yielding varieties" (HYVs).
HYVs differ from normal crop varieties in that the seeds are created under controlled off-farm environments. Farmers may be able to gather seeds from this
crop and repeat the cycle a few times, but the quality of the crop declines, and
after a few cycles they need to return to the off-farm seed source. High-yielding

varieties can give higher outputs of desired crop materials than conventional varieties because they respond to increased inputs—primarily fertilizer and water, but also insecticides.

The development of HYV food crops began in Mexico in 1943, using a genetically based strategy intended to be holistic (i.e., taking the whole eco-system into account). The aim was to produce the most "efficient" plan possible for food production. One important aspect of this efficiency was the ability of the plant to respond to high doses of fertilizer. The program built up an extensive "gene bank" from crop varieties around the world.

One of their most important achievements was to cross wheat from Japan with genes for short stature with Mexican and Colombian wheats to obtain the first dwarf HYV wheats (released in 1961). By 1965, these were the most important wheats in Mexico, yielding up to 400 percent more than those of 1950.

The program's second aim was to examine the growing conditions for the various genetic strains and make institutional and infrastructural recommendations. HYV recommendations included provision for irrigation, improved credit, and agrochemicals. The Mexican program succeeded, and the research strategy and methods of the Mexican program were replicated on an international scale in 1960 when the Rockefeller and Ford Foundations established the International Rice Research Institute (IRRI) in the Philippines. After a series of developments, the Green Revolution is now institutionalized at the highest levels worldwide.

Worldwide, the Green Revolution is considered to have occurred between about 1965 and 1978. Before the technologies could spread, national governments had to see the need and develop the means to adapt them to local conditions. Before the end of World War II, agricultural policy in most of the world was an aspect of the trading and economic policies of colonial empires. In the aftermath of the war, the former colonies became independent nations. At the same time, however, the Cold War dominated world political debate and economic relations these newly independent nations faced enormous problems and faced a sharply polarized outside world from which to seek help.

Western "development economists" and communist ideologists alike saw the key to rising per capita income only in industrialization, and they both argued for drawing resources out of agriculture—particularly "traditional" agriculture—in order to attain it. When these policies were implemented alongside policies to improve public health and sanitation, the effect was a rapid increase in population that was not accompanied by a corresponding increase in agricultural output. The result, in the late 1950s and early 1960s, was a crisis in food production. In 1968, the trend and its probable consequences were summed up in Paul Ehrlich's much-cited book, *The Population Bomb*. It was at this point that governments recognized the need for radical improvements in agriculture and turned to the experience with HYVs to meet that need.

Since 1960, every major agricultural country has adopted HYV crops, and world cereal production has outpaced population growth. Even though this is a result of the Green Revolution, it has not been solely a result of the use of HYVs.

Where HYVs were introduced without proper supporting inputs, they commonly resulted in no increase at all. Where they were introduced with proper supporting inputs in a technical sense but not in a social sense, they underperformed. By contrast, where HYVs were introduced with responsive organizational support and where farmers retained organizational autonomy, similar increases in other crops often accompanied higher yields from the HYV crops.

The Indian state of Punjab established an agricultural university on the American Land Grant model to provide agricultural extension and research support to the farmers and to assist the state government in planning and policy formation. The state also initiated an excellent system of farmer-controlled credit cooperatives, but no one tried to control farmers' decisions on what to grow. The result was that while they adopted HYVs, they also intensified fodder production, and by 1988 the Green Revolution had been followed by a White Revolution in the dairy sector, with important nutritional and economic gains.

Before the Green Revolution, the idea that peasant farmers following traditional practices would be important partners in economic modernization was almost unthinkable. Now it is unthinkable to attempt it without them.

CHAPTER 4

Earth, Air, Fire, and Water

Thomas Midgley, the same research chemist who figured out that
lead would enhance engine performance, had more impact on the
atmosphere than any other single organism in earth history.

—John McNeill (1954—)

Across geological time, the Earth has been a dynamic environment of perpetual change, a process marked by resilience and adaptation. The present represents a momentary end point of a continuum by which earth materials have been cycled and the organisms that are dependent on energy provided by the sun and various earth materials have evolved. This will continue far into the geological future. If we take stock of this present moment, however, we discover that humans are imposing a disturbance to this continuum, with consequences for the home planet and for ourselves: for our *ecosystem*. In the following sections we explore the four classical elements, central to the philosophies of many cultures throughout human existence: air, water, earth, and fire.

Air

Air is the name for those gases in the Earth's atmosphere that animals breathe and plants use in photosynthesis. Air is 78.09 percent nitrogen, 20.95 percent oxygen, 0.93 percent argon, 0.039 percent carbon dioxide, and trace amounts of other gases. The atmosphere is vital to life on Earth because it protects from the sun's ultraviolet radiation, warms the Earth, and controls day-night temperature fluctuations.

Air pollution comes in many forms. Any substance held in the atmosphere can be a pollutant if it occurs in harmful or undesirable concentrations for long periods of time. Even natural occurrences, such as volcanoes and forest fires, can cause serious air pollution. Suspended soil, in the form of dust and sand storms, and often with human causes, can also cause problems in communities near sites of wind erosion and desertification. Throughout history, however, the most common air pollutants have been the result of fuel combustion—particularly from the burning of dung, wood, coal, and petroleum products. These pollutants include carbon dioxide, sulfur dioxide, nitrous oxides, and particulate matter (soot). Other pollutants also have caused problems, including toxic substances, especially lead, mercury, and other heavy metals. Some of these pollutants, such as sulfuric acid, created when sulfurous emissions mix with water vapor in the atmosphere, have wreaked havoc on the health of people and ecosystems in some parts of the world for more a hundred years. Additional pollutants can be harmless in small quantities but can pose serious threats when produced in larger amounts, a development that researchers are still studying in relationship to carbon dioxide and other "greenhouse gases" and ozone, a key ingredient in urban smog.

As with other environmental problems, two other important trends—urbanization and industrialization—have worsened air pollution around the globe. As societies moved into larger cities, particularly those with considerable industrial combustion, people around the globe found themselves living with worsening air quality. In more recent decades, however, the most developed nations have seen a new trend in the amelioration of certain types of air pollution, even while other problems have continued to deepen.

Cultural and political differences around the globe have led to distinct environmental movements and regulatory responses to environmental threats. The principle of doing no harm to a neighbor's property, however, has guided air pollution control efforts through most of history. Ancient Hebrew law and Roman codes, for example, prohibited burning materials in such a manner or place that would harm others. This principle became established in British common law, which held that "nuisances" must be abated when damage appeared. The US 1970 Clean Air Act for the first time set national air quality goals, with the Environmental Protection Agency (EPA) overseeing automobile and factory emissions. A major amendment to the Clean Air Act in 1990 required even further improvements. In several key pollution areas, including particulates and heavy metals, the United States and other developed nations have seen remarkable improvements since the early 1970s.

The internationalization of environmental concern and a growing understanding of the health consequences of air pollution led to reforms in many industrial nations, including Japan, which passed a law to control soot and smoke in 1962, and Taiwan, which passed its landmark Air Pollution Control Act in 1975.

In 1974, scientists suggested the possibility of a global threat—the destruction of the ozone layer by human-made chemicals. By the mid-1980s growing evidence of ozone depletion and the creation of an "ozone hole" over Antarctica convinced governments to act swiftly. A weakened ozone layer posed threats to all life on Earth, including human beings, whom scientists predicted would suffer growing numbers of skin cancers from increased ultraviolet exposure. The nations that led in the consumption of CFCs created in 1987 the Montreal protocol, which required immediate and dramatic decreases in CFC production, at least in the developed world. (Nations in the developing world were able to produce and use CFCs until the enactment of the Copenhagen amendment to the Montreal protocol in 1992.)

Used as propellants in aerosol cans and as a coolant, CFCs had gained wide use in industrialized nations, causing some corporations considerable objection to their ban. Still, substitutes were found without the huge expenses that the regulation opponents claimed would come from the ban. Unfortunately, CFCs already produced continue to damage the ozone layer and will continue to do so for decades to come.

The threat from global warming due to increased emissions of carbon dioxide and other gases (discussed in more detail in chapter 6) has yet to gain the same type of urgent attention. By the 1990s, most climate specialists agreed that continued warming would cause ocean levels to rise as the polar ice caps melted, endangering coastal and island communities around the globe. In addition, scientists predicted changing climate patterns, with some areas become much drier, others wetter, and most areas experiencing greater weather extremes, including more intense storms and longer droughts.

In 1997, the UN convened the Conference on Climate Change in Kyoto, Japan. With both European and American support, the conference established the need for developed nations to control emissions of greenhouse gases, particularly carbon dioxide, and to establish reduction goals. The Kyoto Protocol went into effect in 2005 and has been ratified by 192 parties, although not the United States, and Canada withdrew in 2011.

Water

Water's origins on Earth remain a mystery. It may have been released from within the floating rocky debris that first formed the Earth, or it may have arrived little by little in the form of icy meteors striking the dry young planet. Today, twice as much of the planet is covered by water as is covered by land. Ninety-seven percent of all of Earth's water is found in oceans, and 2 percent is located in polar and mountain ice. Of the remainder, 95 percent is located in underground aquifers, and the rest can be found in lakes, inland seas, surface soils, the atmosphere itself, living biomass, and stream channels.

The ironic phrase "as predictable as weather" summarizes the variable nature of water's stocks and flows: wet and dry periods, droughts, and floods all are common, although their timing and magnitude cannot be predicted. The importance of water to the life of humankind cannot be overstated. Water can be roughly divided into freshwater, which we can drink (albeit not always safely), and saltwater. The health of both forms is vital to the health of the planet as a whole.

Freshwater

Although water covers the majority of the planet's surface, less than 2.5 percent is freshwater, most of which is frozen in glaciers or occurs as groundwater, which can be difficult to access. The available freshwater is vulnerable to overuse and contamination, and water scarcity has been a growing concern as populations rise and resources dwindle.

An exponential rate of population growth, of about 1 percent per annum, has seen the world's population increase from 1 billion in 1800 to 7 billion in 2012. According to current projections, the world's population will reach 9 billion by 2050, with a slight slowing due to decreased birthrates. The rapid growth rate has increased demands for food and water, the availability of which are not necessarily distributed evenly over space or time. Water is truly a precious resource.

Water is *the* key element that sustains life: it is essential for a range of biological processes such as photosynthesis and respiration, the transport of solutes (for example, nutrients and glucose in blood; waste products in perspiration, exhaled air, and urine; and nutrients in plants). Its availability therefore is central to human survival and to life on Earth. In addition, access to clean freshwater is also vital for good health and well-being. The WHO/UNICEF Joint Monitoring Programme for Water Supply and Sanitation reported in its *Progress on Drinking Water and Sanitation 2014 Update* that "despite strong overall progress, 748 million people still did not have access to improved drinking water in 2012, 325 million (43 percent) of whom live[d] in sub-Saharan Africa."

Inextricable links between water and food are evident when we consider that approximately 65–70 percent of global water use is associated with food production. Much of that water is used for irrigated agriculture. Unfortunately, there is a limit to the amount of land available for food production, especially irrigated food production. This means, first, that expansion for irrigation will be at a cost to other land uses such as native forests. Second, there will come a point at which no further expansion can take place.

As human populations grow, there are attendant risks not simply to the volume of water available per person, but to the *quality* of water. Water quality is impacted by the addition of contaminants that include sediments from eroding catchments, nutrients and agrichemicals from intensive pastoral and agricultural land uses, heavy metals and other elements derived from parent rock and soils or from human activities such as mining and industry, pathogens, and complex compounds sourced from sewage treatment plants, industry, and intensive land uses. Contamination by any of these pollutants can affect human health and well-being through, for example, the effects of parasites, diarrheal and other waterborne diseases, metal and other toxicities, and reduced access to potable water.

Access to freshwater is not enough for survival; the water must be of a quality appropriate for its use. The scale of issues and the capacity to implement improved conditions represent significant challenges. The changes that are occurring in the human population, as well as changes human activities have imposed, have stimulated discussion about the risks and uncertainties that lie ahead, and the possible strategies to provide for future generations.

Past practices in a changing world have been exploitive. Although water was identified as a renewable resource, it was not understood that it was a *finite* resource, particularly within the context of increasing demands. Now, the same volume has to be shared by an increasingly large population, which has been, and in many cases continues to be, willing to pollute, overuse, and waste. Also, the *use* of water has to be managed at levels that can be sustained over the long term.

We need to develop this resource while protecting the environment. There are two possible routes toward unsustainable water use. First, it can develop in response to variations in the stocks and flows of water that impact on its availability over time and space. Second, it can develop as water demands change in response to population growth, improvements in technology, increasing standards of living and life expectancies, and evolving social values.

Saltwater

About two-thirds of the world's population lives within 60 kilometers of a coast, and almost one-half of the world's cities with more than 1 million people are located near estuaries. Notable examples include New Orleans (Mississippi River), New York (Hudson), Montreal (St. Lawrence Seaway), London (Thames), Istanbul (the Golden Horn), Guangzhou (Pearl River Delta), and Buenos Aires (Rio de la Plata). This settlement pattern is a result of people choosing the oceans instead of agriculture as a source of food and employment. The oceans also have provided access to communication, transportation, and trade. Today, container ships of unprecedented size ply the world's waters, while vast international fishing fleets and oil platforms extract unsustainable quantities of resources from the depths.

The sequence of conditions called the hydrological cycle (water passes from vapor in the atmosphere, through precipitation on land or water surfaces, and back into the atmosphere as a result of evaporation and transpiration) transports and stores chemicals and heat, determines the Earth's climate, and fertilizes and erodes the land. Ocean surface temperatures around the equator may be 30°C or more, decreasing toward the poles, where seawater freezes at −2°C. Below-surface temperature is fairly constant, decreasing to around 0°C in the deep ocean.

Life began in the oceans, but science has incomplete knowledge of marine life forms. About 15,000 species of marine fish are known, but it is estimated that 5,000 species remain to be identified. The estimate of 200,000 ocean floor species of the North Atlantic alone may be low by a factor of three or four. Whereas the open oceans are "blue deserts" due to the lack of nutrients, the continental shelves are home to abundant marine life, and tropical coral reefs are habitats for large numbers of species, both animal and plant.

Marine Resources

Humans use the oceans as highways for transportation, by exploiting their marine life, and by extracting resources on the ocean bottoms. Oceanic transportation is the cheapest and most important way to move goods between the continents, but it imposes severe environmental stress on marine habitats and biodiversity.

Prior to the fifteenth century, the oceans presented a formidable obstacle to contact between continents. Foraging-era (Paleolithic, 2 million–10,000 BCE) migrants spread from Africa and Eurasia to Australia and the Americas by crossing the straits of Torres and Bering. The Polynesian migrations into the Pacific Islands circa 2000 BCE, as well as the Viking migration across the North Atlantic, also testify to early maritime skills. The subsequent voyages of Columbus across the Atlantic from Spain to the Caribbean Sea, however, opened the way for a sustained exchange—causing great environmental impact—with the New World. This exchange of biota—potatoes, peppers, and tomatoes to the Old World; cattle, goats, and sheep (as well as catastrophic diseases such as smallpox) to the New World—is known as the Columbian Exchange.

With the development of three-masted sailing vessels and nautical instruments, global seafaring allowed an exchange of terrestrial plants and animals that had substantial consequences for the recipient countries. Marine-habitat changes followed as ports and bunker areas for loading coal were extended along shorelines and estuaries, and as mud-dredgers changed tidal currents and coastal erosion. In the twentieth century, tanker ships impacted marine ecosystems when they discharged ballast water transported over thousands of kilometers. Ballast water is one of the most serious threats to marine biodiversity and has caused irreparable changes to ecosystems. Introduced species, which have no natural enemies in new environments, can multiply and eradicate original life forms.

Humans have harvested inshore marine environments since earliest historical times. Humans have fished for whales, seals, fish, crustaceans, and algae and have brought seaweeds, salt, sponges, corals, and pearls to consumers. Today, medicines from anticoagulants to muscle relaxants are derived from marine snails.

Beginning in the sixteenth century, thanks to the shipping revolution, whaling and fishing operations were taken to distant islands and continents. As marine life in these distant waters was depleted, the operations became oceanic, first in the Northern Hemisphere and later in the Southern Hemisphere. These operations extinguished some life forms, such as the Steller's sea cow in the Bering Sea, the gray whale in the European and later the American Atlantic, and the Caribbean monk seal.

The early human impacts on pristine ecosystems are believed to have been important to the whole ecosystem, which may have experienced "regime shifts" (situations in which ecosystems are reorganized from one stable state to another) when top predators that controlled ecosystem dynamics were fished out. Today most of the important fish species are exploited to the full or beyond sustainable levels. Many of the most valuable fish stocks of the world are in decline, and some have become locally extinct. The most dramatic case is the Newfoundland cod fishery, which collapsed in 1991, causing not only enormous and possibly irreparable harm to the ecosystem but also the disappearance of the means of existence for many people in Atlantic Canada.

Commercial exploitation of minerals in the ocean bed is only beginning and is expected to increase dramatically. The ocean bed contains sources of energy in the form of oil and natural gas and minerals such as sodium chloride (salt), manganese, titanium, phosphate, and gold. These valuable metals exist in so-called seafloor massive sulfide deposits, which form on the ocean floor when seawater filters through the permeable crust of the ocean into magma-heated rocks below. The current process of dredging the desired material, bringing it to the surface, sorting the valuable, and returning the nonvaluable material to the ocean floor is thought to be harmful to ocean ecosystems.

Tidal and wave power are sources of energy receiving much attention. The United States' northeast coast, New Zealand, the Korean peninsula, and the United Kingdom all have the right combination of coastal frontage and sustained winds to show potential feasible energy production. Industrialized societies have discharged sewage and other waste into the oceans and also have created the phenomenon of oil spills. The explosion of the Deepwater Horizon drilling platform caused the disastrous spill in 2010 in the Gulf of Mexico, causing the United States' worst environmental disaster. The March 2011 earthquake and ensuing tsunami off the coast of Japan (discussed in chapter 5), spewed radioactive material into the ocean, irradiated a large area of northeast Japan, and caused considerable loss of human life.

Ports and Container Ships

Port towns are the nodes of the maritime transport system. The first port towns were well developed in the Mediterranean 3,000 years ago. Port towns provide access to the economic arteries of a country and therefore have been regulated both for fiscal and military purposes. With the increase in shipping in the nineteenth century, port towns demanded more and more labor and sprawled along wharves and quays to become unwieldy entities, congested and polluted.

As coal-fired steamer traffic spread in the second half of the nineteenth century, a new infrastructure of bunker ports and dedicated quay facilities was built. Wind-powered ships continued to defeat the steamships as long as a line of bunker ports did not dot the margins of the seas, but the steamships took possession of more and more sea routes so that by the early twentieth century the slow windjammers to Australia were the last to give in.

By that time, diesel engines were being introduced, and by the 1950s coal was all but given up. At that point, passenger ships lost their edge in the oceanic transportation of people to the airlines, but a new era for the shipping of goods more than compensated for the loss to ship owners. In the first half of the twentieth century, ships were designed to provide optimum cargo facilities and quicker turnaround times in ports. The Argentine meat industry and the Canary Islands banana trade demanded refrigerated ships, and the oil industry gave rise to tanker ships. By the 1960s, a design revolution took place, introducing the all-purpose shipping container (invented by North Carolina truck-driver Malcom McLean), a metal box that could be refrigerated or otherwise modified and that conformed to standard measures and therefore allowed for convenient storage on board. The container ship became the vehicle for the globalization of trade, which severed the links between origin of resource, modification and packaging, and consumption.

To achieve optimum handling, the once-prolific system of ocean ports has been minimized to a system of a few world ports that are the nodes of a few big container lines. The four largest shipping ports in the world as of 2011 were Singapore, Shanghai, Hong Kong, and Shenzhen, China. Servicing the system are a number of feeder lines from lesser ports and a prolific number of trucking services that ensure that the individual container is brought to its final destination.

The environmental impact of the globalized container system is enormous. Although the system undoubtedly brings rationalization of the economic system, it is dependent on the availability of abundant and cheap energy, which will marginalize, for example, the costs of moving Southeast Asian tiger prawns to Morocco to be peeled by cheap labor before they are transported to Germany to be packaged before they are eventually consumed, for example, in a restaurant in California.

Sea Law and Sea Powers

The oceans and seas have long been governed by law. Hugo Grotius, a Dutch lawyer, historian, and theologian, laid down the first principles for an international law of the seas in his 1609 work *Mare Liberum* (*The Free Sea*). He observed that the sea is an inexhaustible, common property and that all should have open access to it. These principles were adhered to in theory by all major

European naval states and eventually were introduced as the guiding principles for access to all oceans and seas.

Most states claimed dominion over territorial or coastal waters, but commercial ships were usually allowed free passage. The shooting range of a cannon once defined the width of territorial waters, but during the nineteenth century a 3-nautical-mile (5.5-kilometer) limit was increasingly accepted and laid down in international treaties. After World War II, US president Harry Truman claimed wider rights to economic interests on the North American continental shelf against Japan, and Chile and Peru claimed a 200-nautical-mile (370-kilometer) exclusive fishing zone off the coasts against US tuna fishers. Iceland followed soon after with claims to exclude British fishers from Icelandic waters. Oil and fisheries were the main economic motives for these claims.

In 1958, the United Nations called the first International Conference on the Law of the Sea to establish a new consensus on sea law. The conference extended territorial limits to 12 nautical miles (22 kilometers) but failed to settle the issue. A second conference in 1960 made little progress. During the 1960s and 1970s, positions changed dramatically. It became much more evident that the supplies of fish stocks were limited and that depletion was becoming more prevalent. Attempts to manage resources through international bodies were proving ineffective.

Many coastal states, both developed and developing, felt threatened by the large fleets of distant-water states off their coasts. At the same time, the issue of control over the mineral resources in the deep ocean beds raised the demands of developing states for a more equitable distribution of ocean wealth. The third international conference, which lasted from 1974 to 1982, resulted in a convention that is internationally recognized. The main innovation was the declaration of the right of coastal states to a 200-nautical-mile "extended economic zone" (EEZ), which may be claimed by a coastal state for all mineral and living resources. The convention was signed by 157 states (with the United States, United Kingdom, and Germany taking exception to the stipulations on seabed mineral resources). The EEZs represent the largest redistribution of territorial jurisdiction since nineteenth-century colonialism.

The choice of 200 nautical miles as a limit for jurisdiction has no relevance to ecosystems or indeed to the distribution of mineral wealth but is simply a result of international negotiations. Whatever the imperfections of the convention, it has provided coastal states with the authority to manage the resources within their zone. The short history of EEZs shows that they may be implemented to promote conservation interests in addition to the national economic interests for which they were designed.

As of 2011, less than 1 percent of the oceans were protected even by limited restrictions. The oceans and seas are still subject to open access and unrestricted human practices in most regions of the world, and the underwater world remains the last frontier, still to a large degree unexplored by humans.

Earth

From the vastness of the world's lakes and oceans we turn to the earth below our feet. Few subjects are more central to the environment than soil. Soil is the top 1 or 2 meters of the planet's surface, which is tremendously abundant with life. It is the ultimate ecosystem, combining all the elements of other ecosystems and carrying out the critical function of decomposition. Because soil is the geological layer and ecosystem that humans contact first, it bears the strong imprint of humanity. Today humans cause more geomorphic change than any other single Earth surface process (e.g., rivers, wind, and glaciers), with the largest agent of landscape change being soil erosion.

In some senses soil is simply the cover of mineral and organic aggregates that blankets much of the continental surface of the Earth. Most of the large expanses of glaciers, shifting desert sands, and rock pavements do not grow plants, however, and are not soils. Soil has texture (sand, silt, and clay particles), structure (how these textures adhere together in aggregates called "peds"), organic matter (dead plant and animal material), gases and water in micropores and macropores, and a myriad of life forms that parallel aboveground ecosystems but with more emphasis on decomposers. These soil ecosystems are still among the least understood in nature.

Soil is a medium of growth for nearly all plants on Earth. Soil fertility is thus vital, and many characteristics contribute to soil fertility, including nutrient availability, lack of toxic elements, texture, structure, and a healthy microbial ecosystem. Plants need six macronutrients in higher quantities from soil: nitrogen, phosphorus, potassium, calcium, magnesium, and sulfur. Plants also need eight micronutrients at smaller quantities: iron, manganese, copper, zinc, boron, molybdenum, chlorine, and nickel. Within the soil, clays and organic matter store the soil nutrients.

The causes of soil degradation are numerous, although one obvious cause is land use: plowing, irrigation, cultivation of marginal lands, and toxic chemical use and oil exploitation. Greater chemical use, especially of nitrogen fertilizer, has led to higher yields. But large areas of polluted soils have also developed, especially from metal smelting, petroleum, and a stew of other toxic wastes.

The expansion of these processes occurred especially after World War II and spread to the developing world with heavy industry in the 1970s.

Soil erosion has occurred over history in three great waves. The first wave started as humans expanded out of their floodplains in Bronze Age China, the Middle East, and south Asia, about 2000 BCE. The second wave started with the fifteenth-century European expansion around the world, pioneering new lands that Old World farmers had never before encountered. The third wave started after World War II with the tremendous expansion of agriculture onto marginal lands, especially in the tropics, where intense rain and steep slopes combined to produce some of the world's highest rates of erosion. The three waves were similar in one respect: pioneer farmers moved onto lands that they did not understand and that were usually more erodible than lands with which the farmers were familiar.

Fire

One issue of great importance in the human-nature relationship from early times onward is the possession and control of fire. Cooking with fire widened the diets of hunter-gatherers, and mastering fire beyond the hearth helped hunters to capture prey and scare away predators.

Agriculture expanded the realm of fire. Anthropogenic fire was only as powerful as the fuels that fed it. Cultivators could, within bounds, make and break biomass to fashion fuel, although they could not evade the cycles of growth and decay that oversaw how much living biomass was available for use. Humans could hope to transcend this profound cycle only if they could tap another source of combustibles. With fossil biomass, they did precisely that. The combustion of coal, gas, and petroleum launched a new epoch in the Earth's fire history, the era of industrial fire.

Today, the trends in fire are a boom in industrial combustion, a revival of natural fire, and a collapse in anthropogenic burning. How these interact, however, is the vexing three-body problem of fire science. Overall, the Earth seems to suffer a vast maldistribution of burning. There are places, mostly in the developing world, that have too much fire, and places, largely in the developed world, that have too little fire. Along exurban (beyond city suburbs) frontiers, there are also volatile mixed fires. There is too much of the wrong kind of fire and too little of the right. Probably the planet has too much combustion and too little fire. The crisis over global warming is, finally, a crisis over combustion.

CHAPTER 5

Energy Realities and Climate Change

The world is too much with us; late and soon,
Getting and spending, we lay waste our powers;—
Little we see in Nature that is ours;
We have given our hearts away, a sordid boon!

—*William Wordsworth (1770–1850)*

From fire we turn to the issue of energy. Prior to the 1970s, most of the industrialized world did not much consider the amount of energy used or the price paid for it. This was especially true in the United States, which had undergone a period of unprecedented prosperity. Americans were enamored with new household items such as televisions and dishwashers. These items were sold as the epitome of modern convenience, and no one seemed to worry about the amount of energy they consumed. The situation soon changed.

In 1973, Arab oil-exporting countries declared an oil embargo against the United States, Canada, Western Europe, and Japan. The United States' decision to supply the Israeli military with arms during the Yom Kippur War between Syria and Egypt spurred the embargo. Access to oil was diminished, and the price rose. The embargo had a widespread effect on the economies of the United States and Europe. State and federal governments called for conservation by asking businesses and individuals to alter their energy habits. For the first time in history, the United States was not in control of its energy supply, and conservation was the call to arms.

Liquid and gaseous petroleum remain the foundation of our economies and our lives. The issue is not the point at which oil actually runs out, but rather the relation between supply and potential demand. Barring a massive worldwide recession, demand will continue to increase as human populations, petroleum-based agriculture, and economies (especially in Asia) continue to grow.

Humans need to adjust to this new energy reality. Generating a new suite of energy technologies would enable civilization as we know it to continue to evolve. There is certainly sufficient solar energy available to run human civilization. There are three general types of energy investment that we can make. The first is an investment in obtaining energy itself, the second is an investment in maintaining and replacing existing infrastructure, and the third is discretionary expansion. (A related category is investing in what is known as "degrowth," or changing patterns of society that generate well-being but use less energy, as discussed at more length in chapter 6.) Technological improvements, if indeed they are possible, are unlikely to bring back the low investments in energy to which we have grown accustomed.

The main problem that we face is a consequence of the "best first" principle. This is, quite simply, the characteristic of humans to use the highest quality resources first, such as timber, fish, soil, copper ore, and fossil fuels. We have been exploiting fossil fuels for a long time. Our means of finding oil have become less efficient, increasing the amount of energy we invest in seeking and exploiting it.

Several vital resources we take from the ground appear to exist in large supply when matched with current consumption, but there are predictions— and denials—that peaks will occur within the next few decades. Whereas the concept of "peaks" was once central in the discussion of energy resources, it has become difficult to discuss them today. The discussion has shifted to new methods of converting fossil fuels, and especially the idea of "extreme" or not-so-easy oil exploited through methods such as fracking and tar sand treatment. It was due to fracking (the popular term for "hydraulic fracturing") and tar sand treatment that North America became the world's leading oil producer in 2013. Accessing these petroleum supplies that require inputs of technology, money, and time demonstrates the actual scarcity that we face. In the future, however, we will more than likely see a return to the "peak" concept, because these new methods represent a temporary bubble in supply that threatens to make the public think we have plenty.

Renewable Energy: A Short Overview

The term *renewable energy* refers to any naturally occurring and abundant energy source —for example, solar, wind, hydro, biomass, tidal, and geothermal. Although these forms of energy are appealing because many forms of renewable energy are available and do not rely on fossil fuels (and therefore are considered "cleaner"), they are currently more expensive for producers and consumers. Putting a price on carbon emissions has helped level the field between renewables and fossil fuels. The nuclear industry has tried to promote nuclear energy as renewable (on the grounds that it creates minimal greenhouse gases), although not successfully. In 2009, the International Renewable Energy Agency announced that it would not support nuclear energy production programs among its 140 member nations. (Uranium occurs naturally, although, like most materials found in the Earth, it requires mining and processing, not to mention waste disposal). The various renewable energy technologies existing today tend to evolve rapidly.

Hydropower

From as early as the first century BCE, humans have been using water to generate energy. Hydropower is the most widespread renewable source for the world's electricity supply, and even though it is a "mature" (i.e., developed) technology, it has considerable untapped potential. Since the early days of waterwheels and other innovations, human ingenuity has developed new, more efficient energy-production methods—from turbine-driven generators and thermoelectric or hydroelectric power to tidal power and biofuels. One constant remains: water is an essential element in energy generation.

Hydropower has social, economic, and environmental advantages and disadvantages, but the average cost of electricity production by hydropower makes it one of the more "realistic" choices (under current technology) for large-scale renewable energy generation. Most twenty-first-century hydroelectric power plants consist of a reservoir for holding water, a dam that can be opened or closed to control water flow, and a power plant that generates electricity as the water flows through the turbines that spin a generator. Hydropower's major drawback is its virtual irreversibility: once a dam is put in place on a river, with all its possible environmental consequences, it is extraordinarily difficult to remove. Efforts are underway worldwide, however, to remove them.

Tidal and Wave Energy

Tidal and wave energy, or marine energy, comes from the water's motion, which is caused by the wind and by gravitational interactions between the water, Earth, its moon, and the sun. It can be considered a subcategory of hydropower. Whether the various technologies, methods, and devices can generate energy from water remains a subject of ongoing research. There is great power in the oceans and the seas of the world, yet it is not so clear whether such energy can be harnessed in a cost-effective manner. The challenge of wave energy extraction is to "tune" a wave energy device to best capture the considerable—and variable—energy that comes from waves. In general, the best locations for wave energy are near land, where the ocean remains deep and the prevailing wind can transfer energy for thousands of kilometers. Good examples are the western coasts of Spain and Portugal, and the Outer Hebrides off the northwest coast of Scotland.

Geothermal Energy

Derived from the natural heat found within the Earth, geothermal energy is a direct heating source and can generate electricity. The Romans used geothermal energy in the spa town of Bath, England (known at the time as Aqua Solis), to heat their buildings. It is clean, renewable, and abundant. One common misconception about geothermal energy is that it is available only in tectonically active places like Iceland or Yellowstone. In fact, modern geothermal energy technology taps into the difference in temperatures between the surface of the Earth and the temperature underground.

The world's first geothermal electricity generation began at Italy's Larderello (in Tuscany) field in 1902. New Zealand's Wairakei was added in 1958. The Geysers Geothermal Field in northern California came online in 1960, and Mexico's Cerro Prieto in 1970. All of these fields tapped high-temperature vapor that could be used directly for electricity generation. Post-1970 diffusion of geothermal generation technology resulted in construction of new capacities in about a dozen countries. The number of nations using geothermal heat is growing constantly and it is now an option for heating homes in many communities.

Hydrogen and Fuel Cells

Hydrogen and fuel cells have developed since the late-twentieth century as promising means of transitioning to a more sustainable economy. Hydrogen is a nonpolluting energy carrier that can be produced from almost all energy sources; fuel cells are devices that generate energy and heat using hydrogen or other fuels. They are clean, silent, and more efficient than combustion systems.

Hydrogen is not an energy source, and so does not technically belong in this section. Like electricity or steam, it is an energy *carrier* that has to be produced from primary energy sources. Primary energy sources such as fossil, nuclear, and renewable (e.g., solar, wind, biomass) resources are converted to fuels and electricity that consumers use to generate heat, light, and power. Hydrogen has the highest energy content per unit of mass of any known fuel. The use of hydrogen as a transportation fuel has emerged as an active area of research, development, and demonstration. As a fuel, hydrogen can be used in many types of engines, such as internal combustion engines, turbines, and a variety of fuel cells. Fuel-cell-powered vehicles' advantages are emission-free operation and higher efficiencies (by a factor of about two) compared with internal combustion engines.

Bioenergy

Bioenergy is any energy or fuel used to produce energy derived from either biomass or plant matter such as algae, which is being used more for such purposes. Derived from natural resources, bioenergy is a prominent renewable energy source throughout the world. In its most common form, it is wood used for cooking and heating; more complex examples include thermochemical conversions that produce biofuels. Although relevant technologies exist (and continue to be developed), many are not yet cost-effective, and questions remain about the industry's long-term environmental impact. Growing plants for fuel rather than food has led to discussion, and even opposition from members of the environmental justice movement.

In many parts of the world, bioenergy remains the dominant energy source for heating and cooking, but bioenergy is much more than wood used in the home or liquid biofuels like ethanol. Bioenergy production includes biogas and electricity generation from such sources as food-processing wastes and the anaerobic digestion of cow manure. (Anaerobic bacteria digest matter in the absence of oxygen, yielding a biogas rich in methane and carbon dioxide.) In addition to creating jobs and reviving local economies, proponents of agriculture and forestry see bioenergy as a tool to protect productive working landscapes and to provide new markets for these volatile industries.

Solar Energy

People have been harnessing the sun's energy for thousands of years. Salt production by evaporation in arid regions such as Bolivia and the Mediterranean is one use of the sun's rays. In any given hour, the Earth receives more energy from the sun than humanity uses in an entire year. The vast majority of solar energy comes from solar thermal systems that use the sun's energy to heat water either for direct use or to generate electricity. Photovoltaic technology, which produces electricity from specific light frequencies of the sun, produces less solar power. Only a fraction of this energy is usable due to limits in available surface area and technology inefficiency, yet solar power has the potential to meet many of humanity's energy needs.

Solar thermal power can also be used in what is called a solar convection tower, or solar updraft tower. A greenhouse with a tall chimney in the middle covers a large area. Sunlight warms the air in the greenhouse, and the chimney draws the warmed air to the middle. Turbines located at the base of the chimney or in the chimney itself produce electricity as the air rushes by them.

An alternate approach that is being studied would use orbiting satellites as large solar arrays, transmitting energy to the surface of the Earth using low-intensity microwave radiation. The most significant advantage for this technology would be the ability to produce electricity at all times of day in all weather conditions.

Wind Energy

Humans have harnessed wind energy for millennia. Originally used to propel boats with sails, it is now used increasingly to generate electricity with wind turbines. The first written record of windmills comes a millennium after the first mention of water wheels: the Iranian historian and geographer al-Masudi reported in 947 CE the use of simple vertical-shaft windmills in Seistan (in today's eastern Iran) to raise water for irrigating gardens. The first European record comes from the closing decades of the twelfth century.

Wind energy is renewable and emissions-free, but its use as a significant source of electricity faces challenges. Only a small portion of incoming solar radiation (less than 2 percent) powers the atmospheric motion of the Earth. Wind frequencies and velocities on the Earth's surface range from prolonged spells of calm to episodes of violent cyclonic flows (from rainstorms, tornadoes, and hurricanes).

Wind energy supplies a small percentage of energy for electricity generation, although that percentage is growing. As a renewable source with no carbon emissions, wind energy has provoked global interest. The high cost of wind turbines, the difficulty of reaching demand centers, its relative unpredictability, and the variation in actual wind speed (not the theoretical capacity) are its primary disadvantages. Large-scale wind turbine proposals also encounter local resistance from residents who worry that their views will be spoiled by ridge-top turbines or that their property will lose value.

Nonrenewable Energy

The world currently runs on mostly nonrenewable energy, which has many people alarmed due to the greenhouse gases burning fossil fuels emit, and the resulting changes to the world's climate that are becoming more and more difficult to ignore or discredit as the scientific evidence mounts. The most common forms of nonrenewable energy are fossil fuels and nuclear.

Fossil Fuels

Fossil fuels are derived from the remains of plants and animals deposited long ago, mostly during the Carboniferous period, roughly 360 to 286 million years ago. They occur on land and under the sea. The most widely used forms are coal and petroleum (which includes oil and natural gas). Peat, a substance somewhere between coal and wood, is also used for fuel. Petroleum is refined to become gasoline, kerosene, jet fuel, and other derivatives. Fossil fuel as a source of energy is readily available for most populations, and current technology makes it cheap. Although fossil fuels are nonrenewable, their availability and the vast current reserves suggest that humans will rely on them for many years to come.

Coal

Coal is an energy and industrial resource used in the power, iron and steel, chemical, and construction materials industries. It is also burned for fuel in daily life. Its use prevented forests from being cut down for firewood and charcoal. The coal industry, however, consumes resources and affects the environment in numerous negative ways. Coal is a natural resource that is both nonrenewable and not recyclable, and its consumption for fuel leads to higher emissions of acidic rain gases, heavy metals, particulates, and carbon dioxide into the Earth's atmosphere. To make the coal industry at least somewhat less polluting than it is, new technologies need to address the emissions and poor resource utilization that occur during mining, transport, and conversion into power and chemicals.

Coal, accounting for about 30 percent of the world's commercial energy supply, is likely to remain a key energy resource into the mid-twenty-first century. The main challenges to the coal industry pertain to utilization efficiency and the emissions of various pollutants during mining, transport, and consumption. Although many advanced technologies have been developed and deployed to address these problems, coal utilization industries still suffer from low efficiency and high pollution.

Coal is used as an energy resource to produce heat and electricity through combustion. In China, more than 50 percent of the coal consumed annually is burned to generate power. Because coal will continue to be an important resource, cleaner coal technologies are required to ensure energy security and increase the coal industry's sustainability. One developing approach is carbon capture and storage, also called carbon capture and sequestration (CCS). This strategy proposes to reduce the carbon dioxide released by burning coal, petroleum, and natural gas and thus stabilize greenhouse gas concentrations.

Coal mining is dirty and dangerous. It disturbs the land's surface and the underground water, which damages or adversely affects a region's commerce and public life. Open-cut mining, which levels whole mountaintops, as is the case in Appalachia, among other places, is especially egregious to the environment. The notion of "clean coal" refers only to cleaner methods of burning coal: there is no "clean" way to mine coal.

Petroleum: Oil and Natural Gas

The word *petroleum* usually refers to natural liquid and gaseous hydrocarbons and includes oil, natural gas, and natural gas liquids. Sometimes the term is used to describe oil alone. Gas is often found associated with oil, although it has other possible sources, including coal beds and organic-rich shale. What we call *oil* is actually a large family of diverse hydrocarbons whose physical and chemical qualities reflect their different origins and, especially, different degrees of natural processing.

Oil and natural gas provide two-thirds of our energy as well as feedstocks for most chemicals, plastics, and other accouterments of modern life. There has been a close correlation between petroleum use and economic activity, and the recent faltering in global energy availability has been attributed as the critical determinant of many financial problems.

Petroleum is a useful resource from which humans derive fuels, plastics, asphalt, nitrogen fertilizers, tires, paint, and a whole host of organic chemicals ranging from aspirin to xylene (used as a solvent). Petroleum possesses unique qualities, including high energy density and transportability. Due to the magnitude of its use in the twentieth and twenty-first centuries, its future supply has become a cause for concern that requires us to examine the relation between potential demand and possible supply.

We can categorize petroleum as *conventional* and *unconventional*. Conventional petroleum refers to liquid and gaseous fuels derived from geologic deposits, usually found and exploited using drill-bit technology, that move to the surface because of their own pressure, by pumping, or by injecting natural gas, water, or (occasionally) other substances into the reservoir. Unconventional petroleum includes shale oil, tar sands, and other bitumens usually mined as solids, as well as coal bed, "tight sand," shale, and certain other methane deposits that are extracted as gas—often after special treatment such as rock fracturing.

Petroleum is probably contemporary civilization's most important resource beyond sunshine, clean water, and soil. Food production, economies, and cultures of nearly all nations are heavily petroleum-dependent. Nearly two-thirds of the energy used to run the world economy and most industrial countries is based on conventional petroleum, with about the same proportion of use as in the mid-twentieth century.

Oil is especially important for the transportation of people and their goods and services, and as fuel for heating, cooking, and industry. It is a critical feedstock for fertilizers, plastics, most chemicals, and a vast array of contemporary products. Less-developed countries are becoming as dependent upon oil as are developed nations, because development and increased petroleum use go hand in hand. Some people say that we live in an information age or a postindustrial age. Many others believe we live in the *age of petroleum*, for petroleum is the foundation of our economy and nearly everything we do.

Nuclear Power

Debate about the use of nuclear technology (and its destructive and constructive capabilities) continues to rage seven decades after the United States used nuclear bombs against Japan to end World War II. Scientists have long viewed nuclear power as one of the better energy options available, although it should be stressed that nuclear power is *not* a renewable form of energy. Nuclear power creates little immediate pollution (although nuclear waste lingers for millennia), and it uses only small amounts of fuel to make massive amounts of power.

When the first nuclear weapons exploded over Japan in 1945, observers all over the world knew that human life had changed in an instant. In the years since, nuclear technology has struggled to define itself as a public good when the public has seemed more inclined to view it as an evil. Its proponents argue that the electricity that nuclear reactors generate has the capacity to power the world more cleanly than any other resource. Opponents are less sure.

As the debate rages, nuclear power has become an important part of international dynamics: it can liberate nations from the need to trade in fossil fuels, allowing them to acquire inexpensive power that might advance society. At the same time, however, African nations elect to receive nuclear waste from France and elsewhere, which could have health implications for an entire region.

The nuclear reaction is a simple process: similar to power generators that use fossil fuel, nuclear plants use thermal energy to turn turbines that generate electricity. The thermal energy comes from nuclear fission, which occurs when a neutron that a uranium nucleus emits then strikes another uranium nucleus, which then emits more neutrons and emits more heat as it breaks apart. If the new neutrons strike other nuclei, self-sustaining chain reactions take place. These chain reactions are the source of the nuclear energy, which then heats water to power the turbines.

Nuclear power became increasingly popular in the 1970s, despite several nuclear power plant accidents and mounting criticism about the technology's safety. In 1979, the United States experienced its first nuclear accident in a residential area outside Harrisburg, Pennsylvania. The accident at Three Mile Island (TMI) nuclear power plant altered the landscape of US power generation. Although TMI involved only a minor release of radioactive gas, panic ripped through the state. Although the international community took notice of the TMI accident, there was not a widespread perception that the accident threatened the world.

The Soviet Union had even greater difficulty with its atomic industry, which was plagued by accidents throughout this era. None, however, compared to the Chernobyl meltdown that occurred in Ukraine (still a part of the Soviet Union at the time) in 1986. Experts estimate that the accident released thirty to forty times more radioactivity than the atomic bombs dropped on Hiroshima and Nagasaki. Hundreds of people died in the months after the accident, and hundreds of thousands of Ukrainians and Russians had to abandon entire cities.

Since the early 1990s, nuclear power has become one of the world's fastest-growing sources of electricity. The international trend abruptly paused in 2011 with the dramatic events at the Fukushima Daiichi reactor in Japan. On 11 March of that year, an earthquake spawned a tsunami whose tides overwhelmed coastal areas of northeastern Japan, including the nuclear reactors at the Fukushima facility. Of its six reactors, three are known to have experienced full meltdowns. The immediate emergency brought confusion and evacuation of a twenty-kilometer area. The Fukushima disaster focused attention on the need to position reactors properly; for many observers, however, the event presented additional evidence that this technology was not suitable for domestic use.

Today, the risk of the next possible accident constrains the industry. The wild card that could restore nuclear power's promise, according to some, is technological innovation. Although engineers may improve their siting of reactors and better prepare for the next tsunami, nuclear energy seems to be stained in the mind of international consumers by its potential—albeit slight—for cataclysm. Given the increased issues of safety and security, however, the cost of new reactors may prove prohibitive for most nations.

The most sustainable energy future likely includes at least some nuclear power. The industry, however, needs to invest in developing technologies that ameliorate the shortcomings that have nagged the utility since it began commercial production. In the energy industry, many continue to believe that nuclear power remains the best hope to power the future, although some experts disagree, pointing to the risks. In nations with scarce supplies of energy resources, nuclear power—even with its related concerns—remains the most affordable alternative and possibly the most realistic energy choice for the future.

Energy is a complicated issue. The previous discussion has been a fairly traditional one, suggesting that the problem with energy in modern society is one of choosing the right mix of supply. Such a "supply-side approach," where the solution to the problem of a slowing economy is to buy more goods, is common in energy discussions. An alternative approach is the Canadian-Czech scientist Vaclav Smil's idea of an "energy regime": the alternative systems of production, conversion, transmission/movement, and end-use that we have to choose from. The energy source is perhaps the least significant of these elements for assessing the environmental impact of energy use. Another complication is how best to quantify the relative strengths and weaknesses of the various available energy sources. What metrics should we use to compare systems? Should we determine an energy system's suitability by its physical efficiency, cost, carbon emissions, total greenhouse gas emissions, power density, energy return on energy invested, biodiversity impact, national security, or ecological footprint? Although there is no easy answer to these questions, interested readers have a wealth of information to choose from to help them decide.

In these first five chapters we have provided a few examples of the pos-sibilities, problems, solutions, and controversies in the human-environment relationship. In the next chapter we shift the focus to several broad issues and broad solutions for creating a sustainable future for the environment and for humankind.

❧ ❧ ❧

CHAPTER 6

Creating a Sustainable Future

Let ours be a time remembered for the awakening of a new reverence
for life, the firm resolve to achieve sustainability, the quickening of
the struggle for justice and peace, and the joyful celebration of life.

—*Earth Charter, UNESCO*

Sustainability as a word may be somewhat new, having emerged in various places in the early 1970s to describe the idea of conserving resources so that future generations can benefit from them without the economy collapsing in the process. The idea of sustainability, however, is not a new one, despite the increasing use of the word to describe everything ecologically oriented. This chapter reflects on those big issues (which encompass many smaller issues) from the perspective of sustainability, and ventures into the future.

Human Population

The relationship between the human population and the environment is not a simple one. As concern over the environmental degradation has mounted, however, the discussion more often than not comes down to a simple equation: more population equals more environmental degradation. Although this is true in many circumstances, it is not always the case.

The origins of agriculture date to about 10,000 years ago. Where agriculture first took root, population growth accelerated somewhat. With the shift to food production and more sedentary ways of life, the chief constraint on population growth—the difficulty of carrying around small children—eased.

The efficiency of agriculture depends on soils, crops, tools, and other factors, but as a general rule it can support ten times the population density that hunting and gathering can. For this and other reasons, agricultural societies spread at the expense of the less populous communities of hunter-gatherers. Even so, reproductive exuberance barely kept up with disease and famine, which every now and then reached catastrophic levels, pruning back the population growth of better times. This, in broad strokes, was the demographic regime of agrarian society from at least 3000 BCE until 1800 CE.

During that time, the world population grew much faster than it had during preagricultural times, although still slowly in comparison to today's growth rates. There were even times when the population declined. On local and regional scales, epidemics and famines produced such catastrophes fairly regularly, normally at least once or twice in every generation. On the global scale, there were at least two great catastrophes, each of which probably brought global population decline. The first was the great pandemic of the fourteenth century known as the Black Death, probably a result of the spread of bubonic plague throughout most of Asia, Europe, northern Africa, and perhaps parts of sub-Saharan Africa. The second came when the population of the Americas was exposed to Eurasian and African diseases in the wake of Christopher Columbus's voyages.

During the eighteenth century, the human population embarked on its current spectacular expansion. In several parts of the world, epidemics and famines started to recede, and death rates fell. During the nineteenth century, world population almost doubled, and then in the twentieth century it almost quadrupled as death rates tumbled. Better sanitation, vaccines, and antibiotics lowered the toll from disease, and much more productive agriculture increased the food supply.

Population and the Environment

The relationship between population and environment is one of mutual interaction. Environmental conditions affect population's trajectory and population growth (or decline) affects the environment.

The environmental conditions that influenced population were climate, disease, and agriculture. Major climate shifts, such as the waxing and waning of ice ages, affected human population by changing the proportion of the Earth that was habitable and by changing the biological productivity of the parts not covered with ice. The onset of the last Ice Age reduced human population, and its end encouraged population growth.

In most circumstances, population growth has brought accelerated environmental change and continues to do so. In the context of the last half-century, when the population growth reached its maximum rate, the role of population has probably been greater than in previous times. Cropland has increased by one-third since 1950, a process population growth has driven. The proportion of land occupied by roads and buildings has grown in step with, and chiefly because of, population growth. The transformation of habitats both by natural forces (including hurricanes and volcanoes) and human action (including deforestation and the extension of cropland, pastureland, and developed land), has put heightened pressure on many species, especially in tropical forests.

Very often the growth in the human population is directly linked to the growth in the severity and diversity of environmental issues. Is there a point where the earth's resources can no longer support the world's growing population? In recent years, the rate of population growth in some countries has slowed, which has prompted much discussion about implications. For example, there are concerns about the changes in age and class distributions, and the effects of having an aging population. In the future, there will need to be much discussion about the social, economic, political as well as ecological implications of the population and the environment.

Waste: Unused Resources

Historically, the vast amount of waste created by nearly every member of a developed country was confined to large metropolitan areas, and the idea of waste was foreign to most people. Up to the twentieth century, little was wasted, due to poverty or simple thrift. Every item once spent for its primary use was reused for something else, mostly because the majority of goods were not industrially produced. Also, items that were made of natural products such as wood, metal, or cotton tended to last longer and could be fashioned into another product. In rural areas or small towns, merchants would purchase spent textiles for quilts or mattress stuffing and bones left over from cooking to be fashioned into tools. Plastic was not available until later in the twentieth century, and until the end of the nineteenth and early twentieth centuries wastebaskets were not a fixture in most homes.

It was not until the early to mid-twentieth century that industrial production made it possible to create goods that were cheap to make, cheap to purchase, and cheap to replace. As the median household income began to rise, so did the desire to increase one's material goods as a sign of wealth.

The creation of vast amounts of waste and its associated problems are directly related to consumerism in developed countries and economies. The disposal of this waste has many connections to issues of social and environmental justice around the world. The production, storage, and disposal of waste results in many environmental, economic, and social negatives, but also provides many opportunities such as reuse and recycling as economic and social positives.

The framework for evaluating waste streams is known as the "waste hierarchy," referred to as "reduce, reuse, recycle." The first step in the waste hierarchy is "reduce," that is, to reduce consumption of certain items or resources, an idea related to resource conservation. Preventing waste is the most effective way to conserve resources because no energy is consumed. If, for example, consumers reduced the number of grocery plastic bags they use, then petroleum does not have to be extracted and refined and the bag does not become waste.

Next comes "reuse." By reusing items or by purchasing durable items, one is able to sequester the energy placed in that item over its life cycle. Reusing the plastic grocery bag, one prevents energy being consumed to make the new bag. Lowest on the scale is "recycle." Recycling provides many benefits and is the best way to handle material that would otherwise be placed in a landfill or end up as litter. Despite the benefits of recycling, recycling material still involves many energy inputs to collect and process material. It takes much energy to collect and reprocess the plastic grocery bag into another material. The preferred option, not to create the waste in the first place, saves the greatest amount of resources into the future.

Unregulated Waste, E-Waste, and Environmental Justice

Some categories of waste are, thus far, unregulated and until recently have slipped through the cracks of diversion systems. Unregulated waste tends to follow the path of least resistance and ends as a liability far beyond its perceived end life on the curb or in the landfill. These waste streams can be difficult to manage, and their mismanagement has consequences beyond the implications of overconsumption and overflowing landfills. The physical, ethical, and environmental dimensions of unregulated waste can best be seen in the paths of electronic waste (e-waste) and food waste. The solutions to both of these problems lie within the relative simplicity of the waste hierarchy and are likewise illustrative of how a liability can be refashioned into a benefit.

E-waste comprises outdated or unserviceable electronic items such as computers, mobile devices, televisions, remote controls, and music players. A single modern electronic device may contain more than one hundred different elements, compounds, and alloys; older-styled picture-tube monitors may contain more than 3 kilograms of lead, a neurotoxin that causes serious medical conditions. Small quantities of precious metals such as gold and platinum are also contained in most electronic devices, as are rare metals such as tantalum (used in capacitors) and europium (used in video monitors). The production of tantalum from mines in central Africa has contributed to civil conflict in the region.

The raw materials associated with manufacturing these electronic devices are in high demand because recycling levels for the major components are low. The low cost of raw materials, such as the plastics used to make these components, provides little incentive for manufacturers to encourage the recycling of e-waste, although socially marginalized artisanal miners in developing countries are motivated to recover the small quantities of gold and other precious metals from recycled electronic parts.

The environmental impact and the effects on human health associated with the acquisition of raw materials used in popular electronics have been difficult to quantify. The costs therefore have not been adequately reflected in the commercial value of electronics.

E-waste provides an excellent example of a stream of material that crept up on waste managers, industry, and governments and ballooned into a problem that is only starting to be addressed. Before e-waste became an issue for communities and waste managers, most of these items were stockpiled in closets, garages, or corners of people's homes because no one had an idea of how to handle them. When these items are placed in the landfill, they release the chemicals into the land and water; when they are incinerated, they release chemicals into the air. These toxic effluents then travel to the air and water of the communities that live near the landfill or incinerators. In the United States, the first move was to ban these and other toxic materials from landfills and incinerators. Some states require all electronics to be recycled when the consumer is done with them.

The response to the absolute ban of electronic material from landfills was to place electronic material on the "gray market" to be sold to scrap purchasers and shippers who send it to other countries to be disposed of or processed for the small amounts of copper and other metal. This "toxic trade" results in some of the most serious environmental injustice in the late twentieth and early twenty-first centuries. The destinations for this material lie in the lesser-developed countries in Asia, Africa, and Eastern Europe, where strict environmental regulations do not exist. Motivated by the desire for profit, brokers on this gray market sell the material to collectors, who dump the material into lakes or oceans or place them in huge piles near communities that do not have the resources to fight the dumping of toxic material.

Far too often, poor villagers or children dismantle this material by hand to extract the small amounts of marketable material, exposing themselves to toxins like lead. The remains of dismantled electronic material are heaped near villages or towns, where they continue to leach into the water and soil. Even with mandated recycling for electronic material in the United States and Europe, toxic trade is expected to continue because most communities and governments are not prepared to process the mass amount of electronic material expected by mandated e-waste recycling programs.

Of course, the best way to impact the problem of e-waste is to convince manufacturers to increase the usable life of their products or to convince consumers to keep their devices longer. There are many nonprofit and local organizations that accept electronic material and rebuild and refurbish, most notably, computers to be reused by those who do not need the latest technology or cannot afford it.

Food Waste: Liability to Asset

Food waste provides another illustration of how our wastes can be a liability as well as a potential resource. Food waste is composed of unused raw food and uneaten food already cooked or prepared, and it is the third largest component of the overall waste stream by weight. The packaging of food and processed take-out meals also contributes to the food-waste stream, although it is not directly counted in food-waste statistics. In addition to the amounts of energy and fossil fuels needed to produce and distribute food, wasting food contributes to climate change. When food waste enters the waste stream through a landfill it emits methane, which is far more potent as a greenhouse gas than carbon.

Food is formed from the most basic natural resources; natural decomposition returns the nutrients inherent in plants and animals to the soil. Removing these nutrients from the soil cycle and entombing them in landfills depletes soil tilth and breaks natural cycles upon which food production and all life depend. Farmers long understood the value of composting farm wastes into rich humus to augment soil, and indeed many people compost food scraps and other green wastes in backyards. Some large cities now study, collect, and compost food waste as a mechanism to reduce pressure on landfills, to meet government recycling mandates, and to create an economic and environmental resource from what was considered a stinking nuisance.

Looking to the future, there has to be far more serious efforts to reach a stage of zero waste. In the past, there have been many successful projects that have helped to reduce waste. Is it not possible to build on those successful ventures? Many times have local authorities tentatively agreed to a future date when zero waste should be adopted. Those targets have never been reached. What therefore is need to achieve zero waste?

Climate Change

Human-induced climate change, also referred to as *global change* or *global warming*, is a topic frequently in the news. With reports of extreme weather events such as flooding, droughts, and storms, people have become more aware of the potential risks of climate change. Governments of the world have been under increased pressure to take action to reduce these risks. It seems that little progress has been made, however, despite scientific predictions, international protocols, and the desires of some decision makers and organizations to act.

Ecosystems, including human communities, have always had to adapt to climate change as a natural phenomenon. The use of technologies and unsustainable practices such as reliance on fossil fuels, increased human consumption, and a growing world population have led to a crisis in terms of the influence of human activities on world climate. Production of greenhouse gases, especially carbon dioxide (CO_2) and methane (CH_4), has reached levels never recorded before in the past 160,000 years. Most (around three-quarters) of this additional production comes from human use of fossil fuels in industrialized countries.

History and Causes of Global Warming

The connection between global warming and greenhouse gases, especially CO_2, is not a new concept. The principle is simple: the gas-layered atmosphere serves as a barrier for solar radiation, keeping it trapped at the surface of the planet. The principle is referred to as the *greenhouse effect*. Only in the past thirty years have models been sophisticated enough to link recent human activities and the use of fossil fuels to increased greenhouse gases and global warming.

Examining records from monitoring stations around the world, scientists from the Goddard Institute for Space Studies in the United States were able to demonstrate that the global temperature of the Earth has steadily increased over time. The global mean temperature was 0.6° C warmer in the 1990s than in the 1890s. At the same time, they reported that not only was CO_2 increasing in the atmosphere, but so were other greenhouse gases such as methane (CH_4), nitrous oxide (N_2O), chlorofluorocarbons (CFCs) and their substitutes, hydrofluorocarbons (HFCs) and hydrochlorofluorocarbons (HCFCs).

Natural variation in greenhouse gases is common, and climate records, through analysis of ice cores, clearly show that mean temperatures and CO_2 levels in the atmosphere have been varying over the past 400,000 years. The concern comes from the rapid changes that have occurred over the past 120 years. Normally, under natural conditions, there would be a balance in the global carbon cycle. Deforestation, land degradation, and habitat loss have reduced the capacity of some ecosystems to absorb carbon. With greater gas emissions from human activities, the short-term exchanges and reservoirs are unbalanced, leading to more input of greenhouse gases than sinks (a loss of a permanent reservoir such as trees).

The increase of CO_2 in the atmosphere is caused by fossil-fuel consumption, which humans will rely on for many more years. Deforestation, the other major cause of an increase in greenhouse gas (GHG) emissions, is a global phenomenon in which wood is harvested for various uses (fuel, construction material, paper production) or when land is deforested for agricultural use. In many ways, carbon storage is positive because it helps reduce the amount of GHG in the atmosphere.

It's important, however, not to think that just by planting trees we will solve the problem of GHG emissions. The reason is that forests are part of the large sink for carbon accumulation. Ninety percent of the carbon stored on land is located in forests. In the past 130 years, deforestation has caused a decrease in carbon storage in forests by 38 percent. Most of this decrease has come in tropical forests. Contrary to common beliefs, in the past fifty years northern regions of Europe and America increased their carbon storage by the regeneration of forests and the abandonment of agriculture.

Current data show the important role that industrialized countries can play in reducing greenhouse gas emissions through changes in technology, energy sources, and socioeconomic behaviors. Their levels of emissions are such that without changes, it is difficult to see long-term reduction of greenhouse gases. With the accelerated economic growth of several developing countries, it is predicted that the global levels of greenhouse gases will increase. Unless these countries implement important socioeconomic strategies, they will experience situations similar to those in industrialized counties. Because such data can be controversial and can have incredible impacts on most countries, the challenge has been to have adequate and credible data and assessment to convince decision makers of the threat of global warming.

Global Warming Models and Predictions

The concept of global warming does not imply that the temperature is getting warmer at the same rate all over the surface of the Earth. Although it is on average warmer over time, some regions can experience cooling, whereas other regions can experience warming. Many people, particularly in the United States, focus on relatively short-term weather patterns (such as an abnormally cold winter) and thus deny that human-caused climate change is a reality. Land surfaces, for example, warm at a faster rate than do oceans because oceans can store more heat and can assimilate it better than land surfaces. Variation will also occur across a continent due to air current patterns, latitude, and distance from large bodies of water.

Many factors predict global warming, from air and ocean movements to land cover and human activities. Since the 1950s, scientists have developed models to simulate variation and changes in climate and atmosphere. These sophisticated models determine various scenarios expected across the world. There are still limitations, but improvements of resolution and inclusion of more physical parameters continuously improve. The main challenge is predicting the levels of gas emissions in the future. These levels can also reflect different conditions in terms of population or economic growth, technological change, energy use, and so forth. Because experts cannot know how policies and other factors are going to vary in the future, all of these human parameters are assumptions.

Possible Impacts on Ecosystems and Extreme Events

The complexity of the climatic system makes prediction difficult and uncertain. Recent changes can already be attributed to global warming, however, and scientists can predict for some regions what will likely happen over time. Regions such as sub-Saharan Africa, South America, and Southeast Asia can experience increased heat and thus a decline in subsistence agriculture, leading to desertification. Grasslands of North America and Africa may initially have positive impacts from global warming as the photosynthetic rate increases, but unpredictable climatic conditions can reduce crop production in important agricultural regions. Forests of northern latitudes seem to have benefited from global warming. Warmer temperatures may extend the optimal growing conditions of North American forests farther north. A shorter ice period and the decay of permafrost may drastically change the lives of migrating animals and nomadic human populations of the Arctic.

In the oceans, changes in temperatures can bring variations in sea current patterns and a change in the distribution of many species of fish. In both aquatic and terrestrial ecosystems, these potential changes in species can have huge impacts on global biodiversity and rates of extinction. Another major change may be a rise in ocean levels. Estimates of the increase in sea level range from 1 to 10 centimeters per decade during the next century, as polar ice continues to melt in the warmer climate. Experts do not know if these rates are real, but the rates may have disastrous consequences. Because a large number of people live along the world's coastlines, the number of environmental refugees could reach millions of people per year during the next century.

Impact on human health can be severe. Deaths, starvation, and infectious diseases have been predicted, especially in developing countries, where mitigation and adaptation measures have not yet been developed. Displacement of refugees can degenerate into conflicts for new arable lands. Under such degraded conditions, already the case in several countries, infectious diseases (such as cholera, typhoid, and malaria) can explode due to lack of hygiene and poor living conditions. Infectious diseases currently limited to subtropical and tropical regions may spread northward under higher temperatures. In urban areas, human health can be of greater concern. Heat strokes have been more frequently reported in the past few years. For example, France (a country lacking in air conditioning) experienced a heat wave so severe in August 2003 that nearly 15,000 heat-related deaths were reported, mainly among the elderly.

Global Warming Solutions?

Due to uncertainty and the complexity of the processes behind global warming, international policy actions to redirect current levels of fossil-fuel consumption and socioeconomic behaviors have been mostly futile. In 1992, the Framework Convention on Climate Change (FCCC) was ratified, and the Kyoto Protocol was developed from it in 1997. In 2000, 186 countries had signed the agreement and therefore agreed to reduce or stabilize their greenhouse gas emissions to their 1990 levels by the year 2000. Most countries, however, were unable to reach this target.

Industrialized countries face challenges such as an advanced economy, the level of economic activities, heavy reliance on fossil fuels, great distances between urban centers, and, in some cases, cold climate. Industry lobbyists have tried to avoid changes in policies because they are well aware that this might cause loss of revenues for their industries. With globalization and market competition, politicians are sensitive to any policy that could decrease economic growth. Therefore, other solutions need to be found to reduce greenhouse gas emissions while maintaining economic growth at an optimal level. To do so, some industrialized countries are concentrating on two approaches: reducing greenhouse gas emissions or taking greater quantities of greenhouse gases from the atmosphere through carbon sequestration (in which atmospheric CO_2 is reduced or trapped by planting trees, fossil-fuel conservation, etc.).

Another approach to the reduction of greenhouse gas emissions is to add carbon sinks. Some industrialized countries are planning to use the strategy of reforestation. They believe that if they were capable of helping their own industries or the governments of developing countries to plant trees, the global amount of CO_2 would be reduced through carbon sequestration. The main question related to reforestation is how many trees are needed to counter the impact of global warming. Although this is a complex question, the current 6 billion metric tons of carbon emitted into the atmosphere would require about 1 million square kilometers of forests to balance the system.

The coming decades may be the most crucial ones in the history of humankind. With the world population possibly increasing to 9 billion by 2050, with economic activities tripling, and with the desire of developing countries to reach standards of living similar to those of industrialized countries, slowing global warming might not be an achievable goal for many more decades. But how much longer does the Earth have before irreversible damage to key ecosystems occurs? The future of humanity depends on the will of citizens to change their behaviors and to push others to do the same. In the long term, international policies might be useful but might come too late.

The Coming Age: Uncertain Predictions about Sustainability

Projections suggest that population will rise from its present 7 billion to 10.8 billion by 2100. Slight shifts in mortality or fertility would result in large deviations from that estimate, but it serves as a reference point.

At a glance, the basic needs of nearly 11 billion people might seem relatively easy to meet. Today, cultivable land goes uncultivated, or the most productive forms of cultivation go unused. Despite that slack, present food crops and food sources independent of them (such as fishing and livestock grazing) could adequately feed twice the present world population. A closer look reveals obstacles. Demand for meat and dairy is growing. Unsustainable practices help support present yields, and our practice of diverting harvests away from direct consumption is growing. Fossil fuels support farm operations and the production of pesticides, herbicides, and fertilizers. Competition between food and biofuel extends to uncultivated land.

Looking beyond the basic needs of the world's population, modern amenities place an increasing demand on our resources as they continue to support our lifestyles. The largest consumers are heavy industry, buildings (including space heating), agriculture, and transport, but nearly everything that we do today draws down some resources. If 10.8 billion people are to live in large homes, become modern consumers, and roam the world, they will strain these resources. If that is true, then should there not be discussions about quality of life and standards of living? Lowering standards of living does not mean lowering quality of life.

Most visions of a high-energy future focus on nuclear power but are illusory unless the prevailing reactors change from present designs, which convert to energy less than 1 percent of the original uranium (up to 2 percent with reprocessing of fuel rods). If 10.8 billion people consumed as much energy per capita as residents of the United States, Canada, or Australia in 2011 did, and two-thirds of that came from nuclear power, we would need roughly sixty times as much nuclear generating capacity as at present.

At the opposite pole are degrowth scenarios. Although proponents do not reject all modern technology, they emphasize simple wants and simple solutions, agrarianism and self-reliant communities rather than urbanization, the preservation of nature rather than its subjugation, and crafts and light industry rather than heavy industry. The Indian activist Mohandas Gandhi (1869–1948) advocated such an economic system.

Other degrowth proponents envision a makeover of the economy, with a focus on social justice, local currencies, and a new look at labor laws, among other strategies. As the Degrowth Declaration Barcelona 2010 put it:

An international elite and a "global middle class" are causing havoc to the environment through conspicuous consumption and the excessive appropriation of human and natural resources. Their consumption patterns lead to further environmental and social damage when imitated by the rest of society in a vicious circle of status-seeking through the accumulation of material possessions. While irresponsible financial institutions, multi-national corporations and governments are rightly at the forefront of public criticism, this crisis has deeper structural causes. So-called anti-crisis measures that seek to boost economic growth will worsen inequalities and environmental conditions in the long-run. The illusion of a "debt-fuelled growth," i.e. forcing the economy to grow in order to pay debt, will end in social disaster, passing on economic and ecological debts to future generations and to the poor. A process of degrowth of the world economy is inevitable and will ultimately benefit the environment, but the challenge is how to manage the process so that it is socially equitable at national and global scales.

—*Second International Conference on Economic Degrowth for Ecological Sustainability and Social Equity, 26–29 March 2010, Barcelona*

Many advocate population reduction by programs such as China's one-child policy, which only recently has been relaxed somewhat by the government. Chinese officials estimate that approximately 400 million births have been prevented since the program's introduction in 1978.

Venturing into the future, between nuclear-powered megacities and self-reliant ecovillages lie more moderate visions of a highly efficient use of resources and a large role for renewable energy. Proponents of renewable energy maintain it can take over from fossil fuels at present consumption levels or higher. As for efficiency, the most bullish promoters maintain that renewable energy can achieve both sustainability and a high universal standard of living. Whatever the energy source, the system must become self-sustaining, capable of producing its own replacement parts. Electricity from nuclear or renewable sources could deliver the high temperatures needed to work metals and to grind rocks and sinter them into cement. It could split water into hydrogen that could smelt metals. How much electricity such a system would demand or what it would cost to build and run are open questions.

Emerging but as yet unproven technologies could change prospects for renewable or nuclear energy. By producing gaseous or liquid fuels, for instance, artificial photosynthesis could concentrate and store abundant but dispersed and variable solar energy, thus removing obstacles to solar development. Thorium-fueled molten salt breeder reactors would be less prone to serious accidents than present reactors, their wastes would be comparatively low in long-lived radioisotopes, and they could turn thorium into a vast resource, if they prove economical to run. Finally, practical fusion energy perches on the distant horizon, where it has lingered for the last sixty years.

The transition from fossil fuels will demand time. Deployment of existing technology, for example, the building of reactors, wind farms, and metal works, takes years. Nevertheless, researchers and scientists still study (and imagine) the revolutionary potential of technology to conserve and preserve natural resources.

A start toward sustainable resource consumption has never been more strongly in our interest. According to some experts, market forces alone will push entrepreneurs and consumers toward efficient and sustainable resource use. Other bets are on government action, and subsidies are in place in various countries for everything from renewable and nuclear energy to home insulation and natural resource prospecting. If past predictions are any guide, however, no present forecast of a sustainable future (or transition to it) will be correct.

Genetically Modified Foods

Genetically modified (GM) foods are foods that are, or are made from, organisms that have been modified using biotechnology. Such genetically modified organisms (GMOs) contain alien genes, also called "transgenes," that are taken from plants, animals, bacteria, or viruses, or that were created in a laboratory. Genetic engineers have harnessed the mechanisms adapted by bacteria and viruses to overcome the defenses that nature designed to protect individual genomes (an organism's genetic material) from invasion by foreign DNA. This allows engineers to insert novel genes that confer a commercial advantage for agricultural production. These genes include those for traits such as herbicide tolerance, pest resistance, ability to grow faster and bigger, delayed ripening, longer shelf life, and higher oil content.

Biotechnology is also being used to convert plants and animals into factories for producing drugs and other products. This is referred to as "biopharming." Although there are alternate methods of achieving the same food production goals (methods that pose none of the same risks to human health and the environment that GMOs do), bioengineers producing GM foods claim that such foods offer a safe alternative to agricultural chemicals and are necessary to feed the world's expanding human population. GM foods raise issues related to health and environmental safety, as well as to economics, politics, public policy, and international trade relations.

Looking to the future, these issues must urgently be addressed. They are controversial; anti-GM groups label them "Frankenfoods" while pro-GM groups call them "super crops." Consumers get mixed messages about GM foods, and nations are polarized in their acceptance or rejection of GMOs. As the *Economist* put it in a 2014 article on the state of Vermont's decision to ban GMOs by 2016, "The outlook is unappetising. Food scares are easy to start but hard to stop. GM opponents, like climate-change deniers, are deaf to evidence. And the world's hungry people can't vote in Vermont."

Having now explored several pressing issues for the future of humankind, we turn our attention to another: how humans design the built environment in which they live, work, and play.

Design

Sustainable design ideas—whether in the fields of product design, industrial design, or architecture—promote an infrastructure that mimics the natural world. Rather than spending money to design and build sewage-treatment plants that remove organic matter from wastewater, for instance, designers can let marshes do this work naturally. One way to transition to an integrated whole systems design approach is to study nature. What nature does to combat instability in a particular environment involves an integrated or linked diversity in which flows and cycles connect many species, at all scales.

Sustainability is all about processes that are built upon a complex, interlinked diversity. The same criteria can be applied to guiding the design of buildings and cities. In discussing twenty-first-century design in any field, the three Rs—restoration, regeneration, and resiliency—mean integrating design within a larger context of community and the ecological design of food, water, energy, and recycling systems at every scale. Designers have a huge and significant role in helping to achieve sustainability.

Product and Industrial Design

The terms *product design* and *industrial design* in most cases refer to the same field and are interchangeable. Product and industrial designers work in a way that not only fulfills the functional and ergonomic requirements of the "design brief" but also satisfies more subjective criteria such as personal taste and style. (A design brief is typically a list of market and technical requirements, which in recent years often includes sustainability criteria.)

Consumer-led design goes beyond the idea of meeting human needs and instead seeks to create and stimulate human desires. Product design has always been about creating objects of desire, but in today's context of environmental stress, this aspect of design is treated negatively or with caution.

The concept of design for sustainability first emerged in the 1960s, when several design scholars began to criticize modern and unsustainable development and to suggest alternatives. The second wave emerged in the late 1980s and early 1990s and coincided with the green consumer revolution. Writers such as Paul Burall (1991), Dorothy MacKenzie (1991), and Ezio Manzini (1990) began to call for design to make radical changes. This wave continued to gain momentum toward the end of the 1990s and early 2000s as design for sustainability became more widespread.

Since the 1960s, when Austrian-born US educator and designer Victor Papanek first blamed the design profession for creating wasteful products and customer dissatisfaction, there has been a growing feeling in many environmental circles that the design and manufacture process is responsible for many of the stresses imposed on the planet. In the early 1970s Papanek famously stated,

> There are few professions more harmful than industrial design, but only a very few.... By creating whole new species of permanent garbage to clutter up the landscape, and by choosing materials and processes that pollute the air we breathe, designers have become a dangerous breed.... In this age of mass production when everything must be planned and designed, design has become the most powerful tool with which man shapes his tools and environments (and, by extension, society and himself).
>
> —Victor Papanek, *Design for the real world: Human ecology and social change* [2nd ed.]. Chicago: Academy, 1985, ix.

Papanek's assertion is illustrated by the fact that 80 percent of products are discarded after a single use, and 99 percent of materials used are discarded in the first six weeks, according to the design firm Shot in the Dark. Mainstream product design draws on scarce resources to create and power products that have little or no consideration for their impact on society and the environment. The situation is changing, however. For example, European legislation is focusing on the composition of industrial components and how products are processed at the end of their lives.

Design for sustainability takes into consideration social, environmental, and economic issues. Social issues include usability, socially responsible use, sourcing and designing to address human needs. Environmental issues include appropriate materials selection, reducing energy use, and design for end of life. Economic issues include ensuring that the product will sell in its appropriate market. With a few changes, product and industrial design can shift from being seen as a contributor to global environmental and social problems, to being seen as part of the solution to these societal challenges. A good way to visualize this system is to watch US sustainability proponent Annie Leonard's "The Story of Stuff," easily found online.

Architectural Design

Architectural design is at a critical juncture. Some experts believe that the field needs to transition from "business as usual" to a more place-based approach that reduces environmental destruction and adds to what enlightened business calls "the triple bottom line," where economy, ecology, and social equity all support one another.

The realization that many conventional building forms are not sustainable is not new. Buildings in hotter climates (e.g., in the southern United States) once were designed to maximize shade and incorporate features such as verandas for natural cooling, thus minimizing energy usage. In the 1950s, these features began to be excluded as an increasing number of US homes and commercial buildings were constructed as autonomous units with internal mechanized heating, cooling, and ventilation systems that were no longer integrated with the outdoor environment but served to exclude it. This "environmental fortress" mentality of building modernization, which was normalized with the support of the heating, ventilation, and air conditioning (HVAC) industry, rapidly became a template for building practices in the United States and has since proliferated worldwide, driving up building energy use.

In the 1970s, many architects criticized new building designs that sought to exclude nature rather than embrace and use it. Through the 1980s and 1990s, there was continued support for alternative greener architecture and technologies that would conserve energy, minimize resource use, and work with rather than against the climate. Current adaptive-design debates further address the need to rethink how buildings interact with the natural environment in order to create delightful as well as functional and efficient spaces. More recent discussions also propose employing technologies in smarter new ways, such as using adaptive controls or intelligent building materials that can mimic natural processes.

Building designs throughout history have also reflected changing cultural ideas about what is conducive to human comfort and productivity. A preference in North America today is for detached family homes with private yards rather than more compact forms of living. This requirement for individual space has led to urban sprawl, where extensive systems of roads, automobiles, water pipes, and sanitation networks are required to support resource-intensive suburban lifestyles. Critics say that this form of living is not only wasteful but neglects more positive urban sustainability features such as spaces for community interaction. This rejection of suburbia has led contemporary advocates of sustainable urbanism to support more compact and denser building forms.

Public Transportation

A topic related to design is transportation. In many ways, public transit systems represent design on a massive scale. A well-designed transit system can reduce energy use and greenhouse gas emissions compared with travel by automobiles. Transit systems also have a smaller footprint than highways and so can have a lower impact on natural systems and urban environments. Competition from the automobile is a major challenge for most public transportation systems.

Public transit provides both mobility and access; the world's largest cities could not function in their present form without it. Transit use is growing quickly in China, India, and Latin America, where it serves a burgeoning demand for urban jobs and better housing. In the United States and the European Union, transit use is concentrated in large cities, but many small- and medium-sized cities also rely on transit to meet the basic mobility needs of those who cannot drive and to help reduce congestion and emissions.

Latin America continues to make major investments in transit, and has been at the forefront of innovations in bus systems, especially bus rapid transit (BRT). Curitiba, the capital of Paraná State in southeastern Brazil, pioneered BRT and transit-oriented development (TOD), which has resulted in a higher quality of life for its inhabitants. TOD builds markets for transit in both city and suburban locations by creating high-density areas of mixed-use activity within walking distance of transit stations and stops. Curitiba influenced later efforts in some South American cities such as Bogotá and Santiago, as well as in North American cities including San Francisco, Los Angeles, and Portland, Oregon.

In high- and middle-income countries, capital investments in transit are covered mostly by the government, either by directly paying for them or by facilitating joint transit and land development opportunities where revenues from the latter cover transit costs. State, regional, city, and county governments also contribute to transit finance, and plan and regulate transit operations. The specific mechanisms for providing transit funding, however, vary from country to country. Germany takes one approach, offering block grants to state and local governments for the provision of transit services.

Well-designed and well-operated public transit systems offer important economic and environmental benefits. The economic incentives for implementing transit systems are hard to quantify because transit projects are often related to real estate development. Further, transit use can reduce costs even for those who do not use it. Public transportation can help produce benefits from urban agglomeration. Transit can facilitate the clustering of urban activities and help produce economies of scale, with higher productivity and lower costs.

Financing public transportation, however, is a major issue worldwide. Governments usually finance transit systems' capital costs while a combination of fares, other revenues, and government subsidies pay for many transit systems' operating costs. In the United States, local governments subsidize much of mass transit, often by adopting specific property taxes, sales taxes, or fuel taxes. In several places, notably Hong Kong and Japan, land development around transit stations is a major source of revenue for the transportation system.

Looking towards the future, there has to be awareness that public transit faces three main challenges. First, when transit riders are drawn mostly from travelers who formerly used nonmotorized modes of transportation, such as bicycles, pedicabs, and walking, the environmental impact may be negative (although public transit may offer the new riders significant benefits in terms of time savings, comfort, and accessibility). Second, public transit has to compete with the automobile as the basis of transport services in light of its status as an indicator of middle-class success. The third major challenge to public transit is posed by the suburbanization that is occurring worldwide. Transit is problematic in many suburbs around the world, where low densities and auto-oriented street designs make transit services costly and impractical.

Worldwide, transit operators are challenged to put their finances on stable ground, to provide environmentally sound services, to respond to social needs, and to compete with the automobile. There are hopeful prospects for succeeding on all counts, however. New technologies, new operating strategies, and new urban development approaches offer opportunities for public transit to overcome current difficulties and to grow and thrive from a sustainability perspective.

Community

The questions of how we design our products, our buildings, and our transportation systems take us next to a closely related topic: the notion of community. Almost everyone is in favor of "community" as a concept, even when their activities tend to damage or diminish it in reality. The belief that community is essential to sustainability is also enduring, even though over recent decades, television, the internet, social mobility, and globalization have almost completely changed traditional notions of community life.

While the idea of "living locally" gained prominence at the beginning of the twenty-first century, more effort is needed to develop a modern, global replacement for the traditional forms of community. Three aspects of community are particularly important to our thinking about a sustainable future: social cooperation, small-scale resource management and production, and human connection as a source of happiness and well-being.

Community is directly relevant to sustainability in several ways. Close personal ties create practical options for sustainable living. A sense of community also provides personal satisfactions that can make a less resource-intensive way of life more attractive. In addition, online connections can be powerful tools for influencing government and business, as well as for sharing information and supporting efforts to adopt practices that are less polluting or resource-intensive.

Defining Community

An exact understanding of how human communities work is essential to planning a sustainable future. In the same way that scientists can monitor the Earth's changing climate by measuring such things as changes in ice cores and tree rings, the ability to measure what community is, and what a functioning community can do, is important to the pursuit of sustainability.

For environmentalists, the search for community presents conflicting choices. On the one hand, the idea of living and eating "locally" has become a mantra for individuals and even for corporations. On the other hand, global environmental problems require us to act globally, and even those who lead major environmental organizations often do massive amounts of traveling and often lack local community ties. Some argue that the lack of a sense of community is why it remains so difficult to get individuals and nations to make changes even when they acknowledge that the way things are being done results in environmental damage. This raises the question: is the global community of all humans simply too large to generate strong feelings of membership?

The global community has made the pursuit of community infinitely more complicated than it was in the past. Traditional communities were practical. They evolved to meet economic and security needs as well as to provide social and emotional support. One of the challenges for community builders today is that these aspects of life have become compartmentalized.

A renewed sense of community—whatever that looks like—is part of every scenario for a sustainable future. Whereas not everyone espousing sustainability agrees that "degrowth" is essential—indeed, degrowth cuts against the grain of the entire capitalist system as we understand it—many talk about a world in which progress is no longer measured by economic or gross domestic product (GDP) growth, but by levels of well-being and happiness, and by the well-being of the natural world.

Three Key Aspects of Community

Three aspects of community are particularly important to our thinking about a sustainable future: social cooperation, what is called "bridging social capital" practical matters vital to the environment, and happiness and well-being.

Bridging Social Capital

To make changes at the local, regional, national, and international levels, environmentalists need ways to promote a sense of a collective vision and awareness that all humans share a common future. The creation of what is known as "bridging social capital" is needed for large numbers of people who have different priorities and points of view to agree and cooperate. Social capital is a way to describe the value gained from interpersonal relationships, such as business contacts and friendships. Bridging social capital is bringing people together who do not naturally share a close affinity with one another. Bonding social capital is a way to describe warmer, more intimate connections.

Human beings' relationships have profound and long-lasting consequences for the biosphere. Individual rights and self-fulfillment are central to modern ideas about what constitutes an ideal society and about democratic governance, but when discussing a sustainable future, we talk about the importance of collective endeavor.

People today are more connected technologically, but less connected practically and emotionally than in the past. Because of technology and outsourced services, people are able to do everything without help from their neighbors, whereas for most of human history our very survival depended on good relations with those nearby.

An even greater risk, especially for those focused on online communities, is the ease with which people associate only with those who share the same opinions. That is, they are not bridging social capital (and technology reinforces this). Instead of broadening our thinking through access to the vast trove of information available online, we may end up more polarized than ever. This is a disastrous approach if our aim is to cooperate and compromise to solve environmental challenges.

Real-Life Environmental Matters: Localism

The second key aspect of community building is that community has bearing on real-life environmental matters. A sustainable future requires us to balance opposites such as the rural ideal with urban reality (or the urban ideal with the rural reality), and to restructure the suburban in-between.

A big part of this second aspect, the "real-life" value of community, is the idea of localism. Especially since the worldwide economic downturn that started in 2007, the word "local" has come to have desirable connotations and is often paired with "sustainable" in advertisements and marketing campaigns. The idea that locally produced goods are desirable from an environmental standpoint has a much longer history, however, linked with the idea that small communities producing their own food and energy offer the most desirable way to live. Whether these small communities can be sustained in the long term is a matter of debate.

By realigning economic priorities, the degrowth movement seeks to improve individual well-being, strengthen community resilience, and restore the planet's ecological systems. Reducing overall consumption is a primary way to achieve degrowth.

There are a variety of ways that "living locally" is being practiced in the twenty-first century. Community gardens have sprung up in many neighborhoods, urban and rural, in part as a defense against rising food prices, but also as a place to learn gardening tips and to socialize. Locally sourced food, as well as other products and manufactured goods, reduce transportation costs and climate change impact. They may also allow for more efficient recycling or waste composting, and can also reduce storage costs.

Energy can be generated locally, in individual homes, and in towns and regions. Even water supplies can be locally sourced and managed. Countries like Oman, Morocco, Algeria, Syria, Iraq, and Iran are reviving an ancient method of desert irrigation called *qanats* (underground water tunnels that bring water to the surface using only gravity), which originated some 3,000 years ago in the Achaemenid (Persian) Empire.

Happiness and the Search for the Good Life

The third key aspect of community is developing a vision for the future in which our sense of community becomes a source of happiness and well-being. A fully functioning community provides access to that which is good for health and well-being—such as nutritious food, clean air, and exercise.

The search for community is tied to a search for what the Greeks called *eudaimonia*—"the good life." For the ancient Greeks, the question of how to achieve the good life, which included the notions of virtue, as well as happiness and well-being, was of great importance. The philosophy of natural law defined the good life as that which resulted from living in harmony with nature. This ideal of living in harmony with nature's limits, as well as with other human beings, is one that continues to influence environmentalists, sometimes overtly and sometimes subconsciously. The challenge, however, is converting that ideal into a program of action and implementation that will work in the real world.

The Importance of Place

Until the twenty-first century, most people lived out their entire lives within a very small circle, a familiar place in which they were born, raised children, and died. This constancy created a powerful connection to a particular place and region, and therefore a concern about the land's ability to support future generations. For example, the concept of *laojïa*, which has roots in China's history as an agrarian society, remains an important aspect of social identity for Chinese people, even for those overseas. A person's *laojia* isn't necessarily linked to where he or she grew up and went to school. The word translates as "old home" and means where a person's family came from—the ancestral village. Early Chinese immigrants to the United States, for example, would save and carefully wrap the bones of those who died in the new country until they could be carried back to the village for burial.

It is easy to see how "place" matters in the way we live: humans gravitate to places that seem like home, and create new "home" places with remarkable speed. Creating places that meet humans' emotional and psychological needs is one of the challenges facing planners today.

It is becoming less common for people to settle down in the place where they were raised. The forces that are behind this are varied and often interrelated, spanning environmental change through to economics. Indeed, migration in the modern era is made increasingly possible—and desirable—thanks in part to improved telecommunications that allow people to stay in touch easily with loved ones from afar.

City Planning

The relationship between sustainability and community can be explained in many ways, but the questions that arise are quite simple. What should we be doing to create the kind of beneficial communities that will make sustainable living possible? How can a sense of community reduce our impact on the natural world?

Social reformers and, more recently, environmentalists have seen sustainable communities as taking us closer to nature. This approach that can be seen in the American historian and sociologist Lewis Mumford's (1895–1990) ecological regionalism in the 1920s; the "back to the land" ethos of the 1960s, and the ecovillage movement, which began in the mid-1990s; and in calls for rural smallholding and local self-sufficiency. More recently, there has been a procity perspective, with economists and planners (such as those associated with the "New Urbanism" movement) claiming that an urban future is also a "green" future because people who live in cities are less dependent on cars, have smaller homes, and use resources more efficiently.

Jane Jacobs (1916–2006), an American writer and activist who eventually emigrated to Canada, was an urbanist who was highly critical of the arrogance of those who thought they should determine how people ought to live. She was enthusiastic about casual street life and the kind of informal village-like relationships that she believed naturally develop when people live near one another.

Economics and Human Behavior

Community and economics are intricately linked. Much of economics deals with how communities relate, both to each other and to the rest of the world, natural and otherwise. We generally see the natural world as what is described as "public goods." There is little incentive for an individual to avoid using or polluting them, and there is much incentive to "free ride"—that is, to take advantage of anything available. A strong sense of community—whether in real life or online—can change this. The American political economist Elinor Ostrom (1933–2012) received a Nobel Prize for her research into ways in which people collaborate to protect collective goods. This countered the idea that there will always be a tragedy of the commons, the popular idea that individuals will usually use resources to meet their own interest rather than act collaboratively to meet group interests.

The historian William H. McNeill (1917–2016) wrote in his book *The Human Web* that our future depends on finding new kinds of communities to replace those of the past: "Either the gap between cities and villages will somehow be bridged by renegotiating the terms of symbiosis, and/or differently constructed primary communities will arise to counteract the tangled anonymity of urban life. Religious sects and congregations are the principal candidates for this role." Ironically, therefore, to preserve what we have, we and our successors need to change our ways by learning to live simultaneously in a cosmopolitan web and in various and diverse primary communities.

Looking to the future, this call to live both globally and locally is the challenge of sustainability for every one of us, because our physical and psychological well-being depends on the emotional bonds we are able to form with one another. Three approaches seem particularly promising.

First, we need to focus on "bridging" activities that create bonds (and social capital) between people who are different from one another. Organized meetings of stakeholders—all of whom will be affected by a new development or regulation—are designed to do just that. Bridging social capital is created by those ordinary human activities that bring together people from different social and economic classes such as parent-teacher associations (PTAs), pick-up basketball games, block parties, and volunteer associations. This creates the social capital that gives people an incentive to look at environmental issues from other people's points of view and to work together at finding solutions.

Second, we need to recognize that face-to-face relationships are different in essence from online connections. Sustainability is a real-world challenge, and as valuable as online communications may be, change has to take place in the physical world, and within the particular domains that human beings share. A sustainable future will become more likely when we find ways to get out on the sidewalks, into coffeehouses, and in the woods or on the beach with other people.

Finally, we need to explore the ways in which community can satisfy needs and desires, offering sustainable satisfactions and happiness. In the various efforts to redefine our measures of progress, and once certain basic and vital needs are met, most of what matters to people is generated through their relationships with others. Human health is linked to the quality of relationships and the sense of community. By valuing our connections more, and becoming more comfortable with the give and take of today's complex, global relationships and with the differences of perspective that make life interesting, we can find common ground.

The Future

The growing awareness that shortsighted collective behavior today imperils human well-being tomorrow brings urgency to the challenge of understanding and shaping the future. At the core of the notion of sustainability lies a riveting moral imperative: the responsibility of the living to bequeath an undiminished world to the unborn. This obligation requires present generations to adopt an integrated and interdisciplinary perspective that weighs the long-term implications of contemporary practices and adjusts them accordingly.

We live at the moving boundary between completed and uncompleted time. We cast a double-faced gaze both back toward yesterday and ahead toward tomorrow, reflecting on where we have been and imagining where we are heading. A great cultural shift that began with the Renaissance and reached an apotheosis in the Enlightenment brought heightened faith in human reason, science, and progress.

Following this upheaval, modernity released a powerful set of world-changing forces—rapid technological innovation, market economies, democracy, and law-governed institutions—that set population, production, and consumption on exponential growth curves. The ability to see the whole of the Earth from space generated a huge wave of feeling that something must be done to protect our fragile home. With the human impact on nature growing apace, unbounded demographic and economic expansion set a collision course with the limits of a finite planet.

By the latter decades of the twentieth century, concern spread that the human enterprise more and more compromised the ecosphere's capacity to support life, thereby threatening the long-term prospects for human development. That concern has recently been taken up by young people throughout the world, with such initiatives as the school strikes for the climate movement, which began with the fifteen-year-old activist Greta Thunberg protesting outside of the Swedish parliament.

Sustainability invites us to collectively and self-consciously construct the future: to generate plausible images of the world decades from now, establish collective goals, and adapt current choices and behaviors for the journey. Envisioning global futures poses new challenges to both science and the popular imagination. Of the immense web of possibilities opening into the future, only one strand will crystallize into history through the interplay of unfolding patterns, chance, serendipity, and human choice.

In short, everything we do must be underpinned by sustainability.

Postscript: Next Steps

History, despite its wrenching pain, cannot be unlived,
but if faced with courage, need not be lived again.

—*Maya Angelou (1928–2014)*

Today, nearly half a century after the 1972 United Nations Conference on the Human Environment, it is important to put the efforts to achieve sustainability into the context of time, to take stock of our current situation, and to pay tribute to emerging calls for urgent action.

During the last fifty years, hundreds of meetings (global and national) have taken place to address sustainability. Likewise, hundreds of books and reports have been published about and for sustainability. As early as 1987, the United Nations published *Our Common Future*, also known as the Brundtland Report, which predicted that the "greenhouse effect may by early next century have increased average global temperatures enough to shift agricultural production areas, raise sea levels to flood coastal cities, and disrupt national economies."

What can we learn from these years of efforts to achieve sustainability? The contents of this book (especially chapter 6) show that much has happened and that many new initiatives have developed "tools" to help achieve sustainability. There are now many new organizations that have introduced sustainability policies. Many educational institutes have introduced educational and training programs about and for sustainability. Legislation for sustainability has become common practice throughout the world.

There has been an incremental growth in the number of non-governmental organizations (NGOs), an increasing amount of money and resources directed at environmental issues, and an extraordinary growth in global, national, and local environmental goals. One recent example is the UN 17 Sustainable Development Goals, which sets out global goals for the years 2015 to 2030. Another is the 2018 Intergovernmental Panel on Climate Change (IPCC) Report, presenting the data and parameters of climate change. The World Wildlife Fund (WWF) has also launched another Living Planet Report detailing current trends in global biodiversity and the health of the natural world. As this book goes to press, the United Nations has released its 2019 report on global biodiversity assessing changes in the environment over the past fifty years and offering possible outcomes due to these changes.

Environmental legislation and polices have greatly increased over this time period. Activists like the divestment campaigners, who pressure organizations to end sponsorship and investment in fossil fuel companies, and the pipeline protesters such as those trying to block the Dakota Access oil pipeline, as well as NGOs across the planet contribute to sustainability, nature conservation, and environmental management. But all of that diverse effort appears not to be enough. The sustainability crisis is surely becoming more and more urgent. The question is, does sustainability remains as elusive as ever?

The WWF and United Nations Intergovernmental Science-Policy Platform on Biodiversity and Ecosystem Services (IPBES) have both pointed to catastrophic collapse of species numbers and diversity. The WWF estimates that since 1970, the amount of vertebrate biomass on the planet has fallen by 60 percent; in fresh-water habitats, biomass in neotropical zones has fallen by 96 percent. The United Nations 2019 report on diversity stated that "The health of ecosystems on which we and all other species depend is deteriorating more rapidly than ever. We are eroding the very foundations of our economies, livelihoods, food security, health and quality of life worldwide." Evidence is gathering of a "sixth extinction event," because background rates of species extinction is higher than at any time in the last 250 million years.

This as well as climate change has been caused by current consumption rates, but that rate is about to double. Of that increase, only $2 trillion will take place in the developed economies. People in developing economies will demand a larger portion of the cake. Thus, ecological problems will intensify as the middle- and lower-income countries inevitably catch up.

The sustainability crisis is far more urgent and critical than anyone ever imagined. New and radical solutions are urgently needed. Should we be prioritizing investments in sustainability relative to space research? Should sustainability pervade all sectors of society? Should sustainability be at the core of every business in the world, every committee, and every government? Is more legislation required?

Everyone can help and many are helping. All the contributions to sustainability, no matter how small, make a huge difference. More action is needed, and not just reliance on the goodwill of thousands of people and organizations around the world.

Individual countries can make a massive difference. For example, by 2018, China had installed 184GW of wind power and 174GW of solar power. (In 2019, the United States has 97GW of wind and 67GW of solar.) In March, 2019, the *Financial Times* reported that the grid connected tariff of solar had dipped below that of coal for eleven provinces in China, and wind is expected to do the same. That means that in China, because of scale, renewables are cheaper than coal. This is a tremendous achievement with huge consequences for the whole of human society.

China is also investing billions in high-speed electrical trains, which are displacing air travel at distances of up to about 1,200 km. There are 300,000 charge points in China for electrical cars versus 67,000 in the United States. The country has banned gasoline and diesel cars in major cities such as Xi'an and Zhengzhou, and is building a new capital with no surface roads. Meanwhile, it has built extensive subway systems in twenty-six cities.

On the activism front, 2018 saw the rise of Extinction Rebellion, an international group that uses non-violent disobedience to achieve minimal risk of human extinction and ecological collapse. In 2019, The School Strike for Climate (Fridays for Future, Youth for Climate, Youth Climate Strike) was gaining momentum amongst students all over the world.

In the United States, the Green New Deal proposed by the Democrats in 2019 helped make climate change a major issue in the 2020 presidential campaign. Despite the climate deniers who wield federal political power, it's finally beginning to sink in with the broader public that climate change is the threat of our time.

But it is mostly the young people who are crying out for urgent action. Is it because they know that previous generations have stolen resources from today's younger generation? In another fifty years, will people look back and say "it was the youth of this good earth that brought about sustainability, and achieved sustainable and equitable use of nature and the environment?"

The age of sustainability has indeed arrived.

What You Can Do

Many environmental books end with a few paragraphs about "what you can do," and the list invariably begins with "Write to your legislator." Nowadays, they add "Go to a protest." These are worthwhile things to do, but most people thinking about sustainability also want to know what they can do in their daily lives.

The publisher of this book, Karen Christensen, has been writing about green living since the 1980s in such books as *The Armchair Environmentalist* (New York: Hachette Book Group). She's a believer in the power of personal action, and has provided this list of principles for practicing sustainability.

- Small is beautiful. Small cars, small houses, and small pets have less environmental impact, and are easier to maintain and pay for.
- Don't sweat the small stuff. Do take special care over major purchases like a car or refrigerator. Don't agonize over a plastic bag.
- One step at a time. Turn the thermostat down just one degree, not ten. Then another degree, and another. Go vegetarian for a day or two, then longer.
- Be a leader: if you can afford it, be among the first to adopt new technologies like solar water heater or an electric car. This can help take important innovations into the mainstream.
- Watch the weight: anything heavy takes a lot of energy to ship.
- Imitate nature: choose products and methods made from natural materials that can be reused or that will biodegrade.
- Buy things that have already had one owner.
- Share tools, exotic cookware, even a car with friends or neighbors.

Finally, find ways to stay in touch with the world. It's harder and harder, especially if you live in a city, to keep the environment from becoming an abstraction. Office blocks are window-less, and you may rush from bus to subway without a glance at the sky. Notice the tiny new leaves as they appear in spring. Take time to touch the bark of a tree or listen to birdsong.

Reach out to other people. In this increasingly fragmented and technologically driven time, we need to revalue and hold on to our human contacts. Empowerment comes from hands-on involvement. Keep looking for options and evaluating your choices. And take time to enjoy and learn about the beautiful world we live in.

Sources and Further Reading

CHAPTER 1. HUMANITY'S FIRST STEPS

Gardner, D. (2010). *Future babble: Why expert predictions fail—and why we believe them anyway.* Toronto: McClelland and Stewart.

The Human Origins Program—Smithsonian Institution. Retrieved July 30, 2014 from www.humanorigins.si.edu

Infrared Processing and Analysis Center (IPAC) at California Institute of Technology. (2014). How old is Earth? Retrieved August 8, 2014, from http://coolcosmos.ipac.caltech.edu/ask/56-How-old-is-Earth-

Kolbert, E. (2014). *The sixth extinction: An unnatural history.* New York: Henry Holt and Company.

McNeill, J. R. (2001). *Something new under the sun: An environmental history of the twentieth-century world.* New York: W. W. Norton.

Orlowski, J. (Producer & Director). (2012). *Chasing ice* [Motion picture]. United States: Exposure.

Robbins, P., et al. (2013). *Environment and society: A critical introduction.* 2nd edition. New York: Wiley-Blackwell.

Sawyer, G. J. et al., (2007). *The last human: A guide to twenty-two species of extinct humans.* New Haven: Yale University Press.

Schnellenberger, M., and Nordhaus, T. (2011). *Love your monsters: Postenvironmentalism and the Anthropocene.* Oakland, CA: Breakthrough Institute.

Tattersall, I. (2012). *Masters of the planet: The search for our human origins.* London and New York: Palgrave Macmillan Trade.

Turner, B. L. (Ed.). (1990). *The Earth as transformed by human action: Global and regional changes in the biosphere over the past 300 years.* Cambridge, UK: Cambridge University Press Archive.

Zalasiewicz, J., et al. (2008). Are we now living in the Anthropocene? *GSA Today* (Geological Society of America) *18*, 4–8.

Chapter 2: From Foraging to Growing

Balee, W. (Ed.). (1998). *Advances in historical ecology*. New York: Columbia University Press.

Boserup, E. (1965). *The conditions of agricultural growth*. Chicago: Aldine.

Burney, D. (1996). Historical perspectives on human-assisted biological invasions. *Evolutionary Anthropology, 4,* 216–221.

Callicott, J. B., & Nelson, M. P. (1998). *The great new wilderness debate*. Athens: University of Georgia.

Carney, J. (2001). *Black rice: The African origins of rice cultivation in the Americas*. Cambridge, MA: Harvard University Press.

Crosby, A. (2003). *The Columbian exchange: Biological and cultural consequences of 1492*. Westport, CT: Greenwood Press.

Crosby, A. (2006). *Children of the sun: A history of humanity's unappeasable appetite for energy*. New York: W. W. Norton.

Crumley, C. L. (Ed.). (1994). *Historical ecology*. Santa Fe, NM: School of American Research Press.

Curtin, P. (1993). Disease exchange across the tropical Atlantic. *History and Philosophy of the Life Sciences, 15,* 169–196.

Kirch, P. V. (1994). *The wet and the dry: Irrigation and agricultural intensification in Polynesia*. Chicago: University of Chicago Press.

Krech, S., III. (1999). *The ecological Indian: Myth and history*. New York: W. W. Norton.

Lee, R. B., & Daly, R. (Eds.). (1999). *The Cambridge encyclopedia of hunters and gatherers*. Cambridge, UK: Cambridge University Press.

Dodson, J. R. (1992). *The naive lands: Prehistory and environmental change in Australia and the Southwest Pacific*. London: Longman Cheshire.

Mann, C. (2006). *1491: New revelations of the Americas before Columbus*. 2nd ed. New York: Random House.

McNeill, W. H. (1976). *Plagues and peoples*. Garden City, NJ: Anchor Press.

McNeill, J. R., & McNeill, W. H. (2003). *The human web: A bird's-eye view of human history*. New York: W. W. Norton.

Mooney, H. A., & Hobbs, R. J. (Eds.). (2000). *Invasive species in a changing world*. Washington, DC: Island Press.

Moore, P. D. (2002, April. Baffled over bison. *Nature, 416,* 488–489.

Morris, I. (2011.) *Why the West rules…for now: The patterns of history, and what they reveal about the future*. New York: Farrar, Straus and Giroux.

Nabhan, G. P. (1989). *Enduring seeds*. New York: North Point Press.

Pimentel, D., & Pimentel, M. (Eds.). (1996). *Food, energy, and society*. Boulder: University Press of Colorado.

Rindos, D. (1984). *The origins of agriculture: An evolutionary perspective*. Orlando, FL: Academic Press.

Ruthenberg, H. (1971). *Farming systems in the tropics*. London: Clarendon.

Simmons, I. G. (1996). *The environmental impact of later Mesolithic cultures*. Edinburgh, UK: Edinburgh University Press.

Watson, A. (1983). *Agricultural innovation in the early Islamic world: The diffusion of crops and farming techniques*. Cambridge, UK: Cambridge University Press.

Chapter 3: The Shift to Cities and Industry

Berry, B. J. L. (1982). *Comparative urbanization: Divergent paths in the twentieth century*. Basingstoke, UK: Macmillan.

Cernea, M. (Ed.). (1991). *Putting people first: Sociological variables in rural development*. New York: Oxford University Press.

Chandler, T. (1987). *Four thousand years of urban growth: An historical census*. Lewiston, ME: St. David's University Press.

Cohen, J. E. (1996). *How many people can the Earth support?* New York: W. W. Norton.

Commission of the European Communities. (2005, December 21). Taking sustainable use of resources forward: A thematic strategy on the prevention and recycling of waste. Retrieved August 7, 2014, from eur-lex.europa.eu/LexUriServ/LexUriServ .do?uri=COM:2005:0666:FIN:EN:PDF

Deane, P. (1965). *The first Industrial Revolution*. New York: Cambridge University Press.

Food and Agriculture Organization of the United Nations (FAOSTAT). Retrieved August 7, 2014, from http://faostat3.fao.org/faostat-gateway/go/to/home/E

Jain, H. K. (2010). *The Green Revolution: History, impact and future*. Houston, TX: Studium Press.

Hall, P. (1999). *Cities in civilisation: Culture, innovation and urban order*. London: Phoenix Orion.

Hawkins, G. (2006). *The ethics of waste*. Lanham, MD: Rowman & Littlefield.

Landes, D. S. (2003). *The unbound Prometheus: Technological change and industrial development in Western Europe from 1750 to the present*. (2nd ed.) Cambridge, UK: Cambridge University Press.

Landsberg, H. E. (1981). *The urban climate*. New York: Academic Press.

Lansing, J. S. (1991). *Priests and programmers: Technologies of power in the engineered landscape of Bali*. Princeton, NJ: Princeton University Press.

Leaf, M. J. (1998). *Pragmatism and development: The prospect for pluralism in the Third World*. New York: Bergen and Garvey.

MacDonald, J. M. (2000). *Concentration in agribusiness*. Retrieved August 7, 2014, from http://ageconsearch.umn.edu/bitstream/33421/1/fo00ma01.pdf

McFalls, J. A. (2007). *Population: A lively introduction*. Washington, DC: Population Reference Bureau.

Porter, R. (2002). *The economics of waste*. Washington, DC: Resources for the Future.

Randhawa, M. S. (1980–1986). *A history of agriculture in India*. New Delhi: Indian Council of Agricultural Research.

Royte, E. (2005). *Garbage land: On the secret trail of trash*. New York: Little, Brown.

Sassen, S. (2008). *Territory, authority, rights: From medieval to global assemblages*. Princeton, NJ: Princeton University Press.

Schlager, N., & Lauer, J. (2000). *Science and its times: Understanding the social significance of scientific discovery: Vol. 4. 1799–1800*. Detroit: Gale Group.

Sieferle, R. P. (2001). *The subterranean forest: Energy systems and the Industrial Revolution*. Cambridge, UK: White Horse Press.

Short, J. R., & Kim, Y.-H. (1999). *Globalization and the city*. London: Longman.

Thompson, A. (1973). *The dynamics of the Industrial Revolution*. New York: St. Martin's Press.

United States Environmental Protection Agency (EPA). (2009). Retrieved August 12, 2009, from http://www.epa.gov/epawaste/

Von Tunzelmann, G. N. (1978). *Steam power and British industrialization to 1860*. New York: Oxford University Press.

Wheatley, P. (1971). *The pivot of the four quarters*. Chicago: Aldine.

Chapter 4: Environmental Issues and Solutions

Arand, R. P. (1992). *Origin and development of the law of the sea: History of international law revisited*. The Hague, The Netherlands: Nijhoff.

Cech, T. V. (2003). *Principles of water resources: History, development, management, and policy*. New York: John Wiley & Sons.

Cheney, P., & Sullivan, A. (1997). *Grassfires: Fuel, weather, and fire behaviour*. Collingwood, Canada: CSIRO Publishing.

Clapp, B. W. (1994). *An environmental history of Britain since the Industrial Revolution*. London: Longman.

DeBano, L. F., Neary, D. G., & Ffolliott, P. F. (1998). *Fire's effects on ecosystems*. New York: John Wiley and Sons.

Dewey, S. H. (2000). *Don't breathe the air: Air pollution and US environmental politics, 1945–1970*. College Station: Texas A&M University Press.

Gill, A. M., Groves, R. H., & Noble, I. R. (Eds.). (1981). *Fire and the Australian biota*. Canberra: Australian Academy of Science.

Goudsblom, J. (1992). *Fire and civilization*. London: Penguin Press.

Hilborn, R. (2007, June. Biodiversity loss in the ocean: How bad is it? *Science, 316*, 1281–1282.

Hillel, D. (1991). *Out of the Earth: Civilization and the life of the soil*. Berkeley and Los Angeles: University of California Press.

Houde, E., & Brick, K. H. (2001). *Marine protected areas: Tools for sustaining ocean ecosystems*. Washington DC: National Academy Press.

Kaiser, J. (2004). Wounding Earth's fragile skin. *Science, 304,* 1616–1618.

Levine, J. (1996). *Biomass burning and global change.* Cambridge, MA: MIT Press.

McCormick, J. (1997). *Acid Earth: The politics of acid pollution.* London: Earthscan.

Montgomery, D. R. (2012). *Dirt: The erosion of civilizations.* Oakland: University of California Press.

Myers, R. L. (2005). *Living with fire: Sustaining ecosystems and livelihoods through integrated fire management.* Arlington, VA: The Nature Conservancy.

Pimentel, D., et al. (1995). Environmental and economic costs of soil erosion and conservation benefits. *Science, 267,* 117–123.

Postel, S. (1999). *Pillar of sand: Can the irrigation miracle last?* New York: Norton.

Pyne, S. (2001). *Fire: A brief history.* Seattle: University of Washington Press.

Whelan, R. J. (1995). *The ecology of fire.* Cambridge, UK: Cambridge University Press.

Wikander, O. (Ed.). (2000). *Handbook of ancient water technology.* Leiden, The Netherlands: Brill.

Wirth, J. D. (2000). *Smelter smoke in North America: The politics of transborder pollution.* Lawrence: University Press of Kansas.

United Nations. (2014). Framework Convention on Climate Change. Retrieved August 8, 2014, from www.unfccc.int/kyoto_protocol/items/2830.php

United Nations Food and Agriculture Organization. (2011). *The state of world fisheries and aquaculture 2010.* Rome: Food and Agriculture Organization of the United Nations.

United Nations Food and Agriculture Organization. (2014). *World reference base for soil resources 2014.* Rome: Food and Agriculture Organization of the United Nations.

World Health Organization. (2014). *Health topics—Air pollution.* Retrieved August 8, 2014, from www.who.int/topics/air_pollution/en/

Yaalon, D. H. (2000, September 21). Why soil—and soil science—matters? Millennium essay. *Nature, 407,* 301.

Chapter 5: Energy

BP. (2014). *Statistical review of world energy 2014.* Retrieved August 8, 2014, from http://www.bp.com/en/global/corporate/about-bp/energy-economics/statistical-review-of-world-energy.html

Dahl, C. A. (2004). International energy markets: Understanding pricing, policies and profits. Tulsa, OK: PennWell Books.

Freese, B. (2003). *Coal: A human history.* Cambridge, MA: Perseus.

Greenbang. (2012.) Which countries produce the most wind energy? Retrieved 10 September 2014, from http://www.greenbang.com/which-countries-produce-the-most-wind-energy_21841.html

International Energy Agency. (2014). World energy investment outlook 2014. http://www.iea.org/publications/freepublications/publication/name-86205-en.html

International Geothermal Association. (n.d.). Homepage. Retrieved August 8, 2014, from www.geothermalenergy.org

Krupp, F., & Horn, M. (2008). *Earth, the sequel: The race to reinvent energy and stop global warming.* New York: W. W. Norton.

Listen, R. K., & Rosner, R. (2009). The growth of nuclear power: Drivers and constraints. *Daedalus, 138*(4), 19–30.

Nicholls-Lee, R. F., & Turnock, S. R. (2008). Tidal energy extraction: Renewable, sustainable and predictable. *Science Progress 91*(1), 81–111.

Patel, M. R. (2006). *Wind and solar power systems: Design, analysis, and operation.* (2nd Ed.) Boca Raton, FL: Taylor & Francis.

Pogutz, S. et al, (Eds.). (2009). *Innovation, markets and sustainable energy: The challenge of hydrogen and fuel cells.* Cheltenham, UK: Edward Elgar.

Rosillo-Calle, F., et al. (2007). *The biomass assessment handbook: Bioenergy for a sustainable environment.* London: Earthscan.

Smil, V. (2008). *Energy in nature and society: General energetics of complex systems.* Cambridge, MA: MIT Press.

Tester, J. W., et al. (2012). *Sustainable energy: Choosing among options.* Cambridge, MA: MIT Press.

World Energy Council. (2013). *World energy insight: 2013 survey.* Retrieved August 8, 2014, from http://www.worldenergy.org/publications/2013/world-energy-insight-2013/

Chapter 6. Creating a Sustainable Future

Alexandros, N., & Bruinsma, J. (2012). *World agriculture towards 2030/2050: The 2012 revision.* ESA Working Paper No. 12-03, June 2012. Rome: Food and Agriculture Organization of the United Nations.

Bartlett, D. (2012, June 7). *Enabling buildings to "LEEP" forward.* Retrieved August 9, 2012, from http://www.environmentalleader.com/2012/06/07/enabling-buildings-to-leep-forward/

Birch, E. L., & Wachter, S. M. (Eds.). (2010). *Global urbanization.* Philadelphia: University of Pennsylvania Press.

Boana, C., et al. (2008). Environmentally displaced people. *Forced Migration Policy Briefing 1.* Oxford, UK: Refugee Studies Centre, 7–8.

Chapin, F. S., III, et al. (2009). *Principles of ecosystem stewardship: Resilience-based natural resource management in a changing world.* New York: Springer Verlag.

Cooper, G. (1998). *Air-conditioning America: Engineers and the controlled environment, 1900–1960.* Baltimore: Johns Hopkins University Press.

Dryzek, J. S., et al. (Eds). *The Oxford handbook of climate change and society.* Oxford, UK: Oxford University Press.

McDermott, C. (2007). *Design: The key concepts.* London: Routledge.

Ehrlich, P. (1968). *The population bomb*. New York: Sierra Club / Ballantine Books.

Etzioni, A. (2003). *Communitarianism. The spirit of community: Rights, responsibilities, and the communitarian agenda*. New York: Crown Publishing.

European Commission. (2012). *Environment: Sustainable development*. Retrieved August 7, 2014, from http://ec.europa.eu/environment/eussd/

Grubbs, M. A. (Ed.). (2007). *Conversations with Wendell Berry*. Jackson: University Press of Mississippi.

Hardin, G. (1968, December 13). The tragedy of the commons. *Science, 162* (3859), 1243–1248. Retrieved August 8, 2014, from http://www.cs.wright.edu/~swang/cs409/Hardin.pdf

Heschong, L. (1979). *Thermal delight in architecture*. Cambridge, MA: MIT Press.

Intergovernmental Panel on Climate Change. (2014, October). *Fifth assessment report (AR5)*. Retrieved August 8, 2014, from http://www.ipcc.ch/report/ar5/

John, G.; Clements-Croome, D.; & Jeronimidis, G. (2005). Sustainable building solutions: A review of lessons from the natural world. *Building and Environment, 40*(3), 319–328.

Karembu, M., et al. (2009). *Biotech crops in Africa: The final frontier*. Retrieved August 8, 2014, from http://www.isaaa.org/resources/publications/biotech_crops_in_africa/download/Biotech_Crops_in_Africa-The_Final_Frontier.pdf

Kates, R. W., et al. (2001). Sustainability science. *Science, 641–642*.

Layard, J. (2005). *Happiness: Lessons from a new science*. New York: Penguin.

Leonard, Annie. (2007). *The story of stuff*. Free Range Studios. Retrieved October 30, 2014 from https://www.youtube.com/watch?v=gLBE5QAYXp8

Macy, J. (2012). Center for ecoliteracy: The Great Turning. Retrieved August 8, 2014, from http://www.ecoliteracy.org/essays/great-turning

McGrane, S. (2012, July 17). Commuters pedal to work on their very own superhighway. *New York Times*. Retrieved August 8, 2014, from http://www.nytimes.com/2012/07/18/world/europe/in-denmark-pedaling-to-work-on-a-superhighway.html?_r=2&pagewanted=all

Nearing, S., & Nearing, H. (1990). *The good life: Helen and Scott Nearing's sixty years of self-sufficient living*. New York: Schocken Books, Inc.

Ng, W.-S., & Schipper, L. (2006, December). Rapid motorization in China: Policy options in a world of transport challenges. Retrieved February 24, 2010, from http://www.cleanairnet.org/caiasia/1412/article-71604.html

Norman, D. A. (2004). *Emotional design: Why we love (or hate) everyday things*. New York: Basic Books.

Oldenburg, R. (1999). *The great good place: Cafes, coffee shops, bookstores, bars, hair salons, and other hangouts at the heart of a community*. New York: Marlowe & Company.

Papanek, V. (1985). *Design for the real world: Human ecology and social change*. London: Thames and Hudson.

Putnam, R. (2000). *Bowling alone: The collapse and revival of American community*. New York: Simon & Schuster.

Robin, M-M. (2012). *The world according to Monsanto*. New York: New Press.

Ruddiman, W. F. (2005). *Plows, plagues, and petroleum: How humans took control of climate*. Princeton, NJ: Princeton University Press.

Solnit, R. (2009). *A paradise built in hell: The extraordinary communities that arise in disaster*. New York: Penguin Books.

Somsen, H. (Ed.). (2007). *The regulatory challenge of biotechnology: Human genetics, food and patents*. Cheltenham, UK: Edward Elgar.

Soskolne, C. L., et al. (2007). *Sustaining life on Earth: Environmental and human health through global governance*. Lanham, MD: Lexington Books.

United Nations Framework Convention on Climate Change. (2014). Homepage. Retrieved August 8, 2014, from www.unfccc.int

United Nations Population Division. Retrieved August 7, 2014, from http://www.un.org/en/development/desa/population/

Van der Ryn, S. (2013). *Empathic design*. Washington, DC: Island Press.

Westra, L. (2006). *Environmental justice and the rights of unborn and future generations*. London: Earthscan.

Whiteley, N. (1994). *Design for society*. London: Reaktion Books.

World Wildlife Fund (WWF). *Living Planet Report 2014*. Retrieved October 3, 2014, from www.panda.org

Postscript

Butler, N. (2019, March 11). Wanted: Long-distance high-voltage grids that can link continents. *Financial Times*. Retrieved April 15, 2019, from https://www.ft.com/content/266a2d1c-248f-11e9-b20d-5376ca5216eb

Hill, J. S. (2019, January 23). China installs 44.3 Gigawatts of solar in 2018. Retrieved April 15, 2019, from https://cleantechnica.com/2019/01/23/china-installs-44-3-gigawatts-of-solar-in-2018

WCED. 1987. *Our common future*. Oxford: Oxford University Press.

Wood, H.W. 1985. The United Nations world charter for nature: The developing nations' initiative to establish protection for the environment. *Ecology Law Quarterly*, 12, 977–996.

Acknowledgements

I n 2007, in response to the growing sustainability crisis, the Berkshire
Publishing Group set about preparing a ground-breaking interdisciplinary
resource for twenty-first-century students and professionals. The ten-
volume *Berkshire Encyclopedia of Sustainability* was published between 2008
to 2012 and has been since translated into Chinese.

As we compiled the final volumes, we thought that a "pocket version"
would be a very useful and logical supplement. More than 1,000 contributors
from around the world provided an extraordinary range of information and
perspectives on sustainability-related topics, and without those contributions
this book would not exist. There are too many contributors to mention all by
name, but they include the original staff editor, Bill Siever, and anthropolo-
gist David Levinson, who has written and edited other Berkshire books. John
McNeill gave helpful advice, as did another environmental historian, Leif
Fredrickson.

I also want to acknowledge the authors whose work in the *Encyclopedia of
Sustainability* we drew on: Timothy Beach, Georgetown University; Rebecca
M. Bratspies, City University of New York School of Law; Karen L. Bushaw-
Newton, Northern Virginia Community College, Annandale; Heather
Chappells, Dalhousie University and Saint Mary's University; Norman L.
Christensen Jr., Duke University; Elizabeth Deakin, University of California,
Berkeley; Loretta Annelise Feris, University of Cape Town; John Martin Gillroy,
Lehigh University; Brent M. Haddad, University of California, Santa Cruz;
Matthias Heymann, Aarhus University; Poul Holm, Trinity College, Dublin;
Abhimanyu George Jain, National Law School of India University; Vicky
Lofthouse, Loughborough University; J. R. McNeill, Georgetown University;
Adrian DiCianno Newall, energy consultant, Westfield, New Jersey; Oladele A.
Ogunseitan, University of California, Irvine; Gregory L. Possehl, University of

Pennsylvania; Stephen J. Pyne, Arizona State University; Evandro C. Santos, Jackson State University; Michael D. Sims, independent scholar, Eugene, Oregon; Vaclav Smil, University of Manitoba; David Stradling, University of Cincinnati; U. Rashid Sumaila, University of British Columbia; Petra J. E. M. van Dam, VU University Amsterdam; Jon M. van Dyke, University of Hawaii; Daniel E. Vasey, Divine Word College; Gernot Wagner, Environmental Defense Fund; D. G. Webster, Dartmouth College; and Verena Winiwarter, University of Vienna.

There is one person who not only made the encyclopedia happen, but also contributed greatly to this book (including the What You Can Do section at the end), and that is Karen Christensen. The CEO of Berkshire Publishing Group, Karen has been an advocate, editor, and publisher for environment and sustainability causes since the 1980s. She has also written her own environmental books, including *Home Ecology* in 1989, *Eco Living* in 2000, and *The Armchair Environmentalist* in 2004. She is currently working on a new, definitive handbook on green living and continues to advocate for various environmental causes.

Ian Spellerberg

Author Biography

John Maillard

Ian Spellerberg is professor emeritus at Lincoln University in New Zealand and the author of more than twenty books, including several on the social history of everyday items such as milk cans and matches. He has had a distinguished career focused on sustainability since the 1970s and is one of the editors of the *Berkshire Encyclopedia of Sustainability*. Spellerberg helped to establish one of the first interdisciplinary environmental sciences BSc degree programmes at the University of Southampton, UK, and later led the interdisciplinary Centre for Resource Management (CRM) at Lincoln University. He was active in the establishment of international masters programs in cooperation with BOKU University in Vienna and Goettingen University (Germany). He has jointly published on education for sustainability and has helped to implement university campus sustainability advisory groups. He is an Honorary Fellow of the Environment Institute of Australia and New Zealand.

Index

A Note about the Book Design

In many ways, *What Is Sustainability?* echoes the *Berkshire Encyclopedia of Sustainability*. We are again delighted to use a prairie photograph on the cover. Taken in Iowa by Carl Kurtz, this photo shows fireflies (*Pyractomena borealis*) at night on a section of restored prairie. Although the prairie of the Midwestern United States is a particularly challenging ecosystem to restore, farmers and naturalists like Kurtz have worked closely with scientists to develop methods for doing so. Karen Christensen selected this image because it so vividly shows the ephemeral beauty of the natural world. While this restored habitat may not have the grandeur of the Alps or the drama of a stretch of coastline, it is beautiful in a way anyone can recognize and has symbolic resonance. Its myriad points of light remind us of all the individual and collective efforts to work on behalf of the earth. The fact that even the prairie can be brought back to life shows that a sustainable future is within reach.

We continued to use the Harrington typeface for titles because we love its sweeping curves, and especially the S, which echoes the spiral of the unfurling ferns we have used on the cover. For the text itself, however, we've used a beautiful typeface called Filosofia, inspired by the book *The Perfectionists*, by a friend and neighbor Simon Winchester. The type in *The Perfectionists* was set in a 1996 interpretation of the eighteenth-century Didone serif typeface Bodoni, by the Bratislava-born type designer Zuzana Licko. Like Simon, we especially like its below-the-baseline numbers 3, 4, 5, 7, and 9, its slightly-bulging serifs. We chose it to make a point. Sometimes people think that a sustainable world will be ugly and utilitarian, and we want this book's design to reflect our conviction that sustainability offers the joy and beauty that humans have through the ages found in the natural world.